I0232448

TRANSFORMATION OF THE
TECHNOLOGICAL SOCIETY

Transformation of the Technological Society

Egbert Schuurman

Dordt Press

Cover design by Vaughn Donahue
Layout by Carla Goslinga

Copyright © 2022 Egbert Schuurman

This is an edited translation of Egbert Schuurman's *Tegendraads nadenken over Techniek* (Eburon 2014) by Kim Batteau. Chapter 2 is an edited version of a lecture originally given to a gathering of the student organization Civitas Studiosorum in Fundamento Reformato, held in Utrecht, The Netherlands, in 1996. Chapter 3 is an edited version of a lecture originally given to a gathering of the Association of Reformed Students in Eindhoven, The Netherlands, in 2002. Chapter 4 is an edited version of Schuurman's exaugural lecture, 15 May 2002, at Delft University of Technology. Chapter 5 is Schuurman's exaugural lecture, 20 September 2007, at Wageningen University. Chapter 6 is an expanded special lecture for the student organization, Civitas Studiosorum in Fundamento Reformato of Delft University of Technology, 2013.

All quotations from Scripture are taken from The Holy Bible, New International Version® NIV®. Copyright © 1973, 1978, 1984, 2011 by Biblica, Inc.TM Used by permission. All rights reserved worldwide.

Fragmentary portions of this book may be freely used by those who are interested in sharing the author's insights and observations, so long as the material is not pirated for monetary gain and so long as proper credit is visibly given to the author. Others, and those who wish to use larger sections of text, must seek written permission from the publisher.

ISBN: 978-0-932914-15-6

Printed in the United States of America.

Dordt Press   www.dordt.edu/DCPcatalog
700 7th Street NE
Sioux Center, Iowa 51250

The Library of Congress Cataloging-in-Publication Data is on file with the Library of Congress, Washington, D.C.

Library of Congress Control Number: 2022909861

CONTENTS

FORWARD

In the course of time, I have had the opportunity to lecture and write extensively about the philosophy of technology.

The first chapter of this book was written as an essay in connection with the Dutch Month of Philosophy's theme, *Humanity and Technology* (April 2014).

The following chapters contain various university lectures. Starting in 2000, I gave lectures that culminated in a lecture for university teachers and students at Delft University of Technology, in The Netherlands, in the spring of 2013. Here I focus on the multiple crises of our culture, and the central role of modern technology.

Parenthetically: the term "technology" is commonly used, in English, to refer both to *technological theory* (a field of study at universities) as well as to *technical applications* (like robots). *Technology* in its literal sense (from the Greek *technay*, "art, skill," and *logos*, "speech, reason") is the *scientific study* of technical applications. *Technical applications* are the practical results of such scientific study. In this collection of lectures, I use the terms "technology" and "technical applications" as synonyms, except when I am referring to "technology" as a field of university study.

The title of this book, *Transformation of the Technological Society*, expresses my belief that a critical reassessment of technological advancements is desperately needed – very much against the grain of most approaches. New technology in the realm of the Internet, smart phones, robots, etc., often goes together with soaring high hopes, even messianic expectations, for their success. In reality, things are quite different.

In this volume, I focus attention on the underlying, driving *spirit* of technological development. When we do this, we can deepen and widen our view of what is actually happening around us. We then can get a glimpse of an alternative, promising perspective, giving us a realistic hope about the future.

I certainly share much with other philosophers of technology — we live indeed in the same world and face the same problems — but our views of life differ substantially. I do not at all reject modern technology outright, but rather seek to appreciate its true *value*, and to develop a *responsible perspective* on technical development. I invite you to consider the merits of my alternative vision.

Time is of the essence. It's time to wake up!

Egbert Schuurman
Breukelen, The Netherlands, 2017

CHAPTER 1

HUMANITY AND TECHNOLOGY:
A CHALLENGING HISTORY

Introduction

There has never been such a technically oriented age as ours. Information technology, nanotechnology, biotechnology, and neurotechnology are the cause of the second technological revolution currently taking place. Computers, robots, and cyborgs—a "fusing" of human beings and machines, and genetically modified or manipulated organisms are advancing rapidly. Recently, during the "month of philosophy" in The Netherlands, much attention was given to this phenomenon. But, surprisingly, little attention was given to the cultural, historical background of thinking about humanity and technology. The problems were rather one-sidedly elucidated and evaluated from a philosophical perspective.

An alternative approach, "against-the-grain," from a perspective of a Christian, philosophical cosmology, anthropology, and ethics, is needed. I believe that this kind of an approach does more justice to the history of the relation between technology and philosophy, and challenges us to see the increased human responsibility involved. For Christian politics as well, a special contribution in the public discussion will be on the agenda.

The history of tools

From the very beginning of the history of human culture, human beings have used tools. During many centuries, there was scarcely any development or renewal of the tools at human beings' disposal. We used the techniques made possible by tools to manipulate and give form to Nature in order to reach certain aims.

It is obvious that humanity is able to do more with tools than without them. The scope and the achievements of human work increase in proportion to how developed our set of tools has become. We could say that—with exceptions—human beings objectivize or project certain functions and skills through tools, and thereby strengthen them and increase their scope and effectiveness. Human beings are measurably diminished through this process. In the course of time, especially since the impact of modern scientific, technical management and control, the development of our tools has accelerated considerably. By means of the principles of cybernetics—the science of technical guidance—machines are given the ability first, to organize, and then to give guidance to systems. As a consequence, we have now entered an age of information technology, after the initial phase of material and energy technology.

Through this development of technology, human beings are physically removing themselves more and more from that which is actually happening. Our power and abilities are primarily involved with the initial stages of developing systems which are dominated by scientific-technical developmental methods. It is evident that even at these preliminary stages, the computer is taking over more and more of what human beings used to do. We have approached the threshold of completely new age of development.

The computer is a fundamental, and at the same time fascinating, technical achievement. In the last 40 years, the development of the computer has been overwhelming. Starting with large, unstable, expensive computers, the development has been in the direction of small, stable, fast-working, and cheaper computers. Microcomputers and microchips serve to accelerate the technical developments. This process is continuing in the direction of quantum-computers and bio-computers. All of this means an advance of the technical production process toward complete automatization. In the meantime, 3D printers have come on the scene. Due to these printers, the assembly line is beginning to disappear, and production is more and more individualized and automated. Computer models of desired products lead to manufacturing faster and with astonishingly more accuracy than in the past. The potential for the applications of such 3D printers is almost limitless.

By means of the latest developments, not only is a lot of technical work of many people being replaced, all kinds of cultural activity is being automated. We now speak of "artificial intelligence" when we refer to projections of the "sensory" and "mind" functions of human beings. With the development of industrial and non-industrial robots, efforts are being made, and with success, to transfer functions of human beings to robots, and some functions have yet to be adequately described and formalized.

We speak of a robot when its "mind" is a computer and the information to be processed is derived from both the programs of the software and also from external input, via sensors. Robots then often act through artificial limbs, resulting in effects in the outside world. Further, a robot can often "learn" from its own experience and "improve" itself. Such robots are already able to do much more than the fantasy many imagined being possible. Robots are making great inroads in the operating rooms of hospitals, in homes, doing housekeeping, are replacing soldiers in the army, are flying unmanned airplanes—drones—and are being used extensively to explore the planets of our solar system. Robots can also be deployed in cleaning up that which is too dangerous for humans to come near, such the destroyed nuclear plants at Fukushima. Such robots are not merely "dumb machines" simply carrying out the instructions of computer programs, but are able to adapt to their surroundings and learn from them. They are often dispatched from a distance, but can then "independently" carry on the needed activity at the site of the operation. Through this last development of technology, we human beings can now do much more than we are able to do "by nature."

Explosive technological development

Since the 1980's, many new technical developments have taken place, which have put the relation between human beings and technical applications at the center of attention. Next to the advance of personal computers and the social media, since 1980 we've also seen the rise of biotechnology, nanotechnology, and neurotechnology. Since then, these technical applications have increasingly become part of our daily lives. They are finding their way into the center of our public and private activities in all kinds of ways. The home, the workplace, the sectors of transportation, food production, energy, and entertainment, during free time, in education, all these areas have become parts of one great technical network. We human beings ourselves are being changed at a deep level through our interactions with these modern forms of technology (IJsselstein, 2013).

I predict that the *Google Glass*—glasses which when worn can connect us to the Internet and cyberspace everywhere and at every moment—will also have a considerable impact. This could be positive. For example, if you are forgetful, these glasses can remind you of someone's birthday. How will this affect the way in which we relate to each other? Wouldn't it be easy to be suspicious when someone meets me with his Google Glass on, thinking that he or she might be peering into my personal past via Internet at the same time? If some of this information is sensitive, it could be tricky. Won't meeting each other be overshadowed by this awareness? Will real openness be possible? In any case, it seems clear that our personal encounters will be endangered and even undermined by these new technical applications, as Sherry Turkle (2011) warns us.

Human beings can *themselves* become the object of technical management and control. Whereas in the past such management and control had to do with reality outside our lives, now new technology is being aimed at human beings themselves. When human beings are physically fused with technical applications, we are heading in the direction of what is called cyborgs. "Cyborg" is a term derived from the words *cy*bernetic *org*anism, and is the physical fusion of a human being with a machine. This can go so far that—in the opinion of some—the distinction between a human being and a machine disappears (Haraway, 1991).

In the meantime, nanotechnology has made its appearance. Nanotechnology refers to working with particles on the order of nanometers (a billionth of a meter). Nanotechnology is thus a technical application which manipulates materials at this extremely small scale. Conglomerates of atoms can be constructed ("nanoparticles"), which have completely different characteristics from natural conglomerates of the same materials. They react sometimes much more violently, they cluster together differently, are sometimes difficult to break down, and, just because they are so different, it is not known what they will do when, for example, they enter the human body. Nanotechnology thus implies the construction of a new, artificial "Nature" at the most basic level of the characteristics of matter. This may mean that new and very favorable characteristics will emerge, but also entails the possibility—as yet unknown—of real risks for human beings and the environment. Nanotechnology affects life itself, because it is a technical application by which DNA molecules can be manipulated and people think that they can make living cells. This increases the possibility of gene technology being able to put pieces of hereditary material of one sort into the DNA of another sort. A tiny instrument made at nanoscale—a so-called "nanobot"—can be placed in the body and serve as a mini-laboratory which, for example, internally measures blood values, and, depending on the results, make corrections (Oomen, 2010). Nanotechnology thus affects that which has been called intimate technology (a term of the Dutch Rathenau Institute in 2014).

Even the human mind can be influenced by nanotechnology. Think of how it is possible to implant chips in the human brain, whereby nerve cells grow directly on electrodes. In such neurotransplants, nanotechnology, biotechnology, information technology, neurotechnology, or cognitive technology come together, a merging of technical applications and the sciences which belong to them. Human beings are more and more being equipped with such technology. People say, "Humans are becoming machines, and machines human beings," and "Biology is becoming technology, and technology is becoming biology." This is clearly so in the case of so-called "Deep Brain Stimulation" (DBS). This takes place via a neuro-stimulator, mostly put under the collarbone, which sends electrical impulses to an electrode in the brain. In this way, certain pathological neurological behavior patterns can be changed—for example, in the case of Parkinson's disease—or new behavior can be stimulated, whereby the identity of the person concerned can be changed (Rathenau Institute, 2014).

"Brain-Computer Interface" (BCI) has focused above all on neuro-prostheses, which are geared to restoring damaged hearing, sight, and movement abilities. Thanks to the remarkable cortical plasticity of the brain, signals of implanted prostheses are processed by the brain, so that that which has been damaged or is missing can be healed or restored (Warwick, 2004). Even interfaces between the body's nerve cells and robots are being made. A miniature robot no longer has to be controlled by a computer—this can be done directly by brain cells. We see a comparable, striking example of a mind-machine breakthrough in a reverse situation; having neurons and electrodes grow together allows chips to be placed in the brain, which are then controlled by external signals, or which can be read by an external apparatus and whereby influence on the behavior and intentions of the person involved. Having nerve cells grow together to electrodes is something which now is taking place at many research institutions in The Netherlands. Thanks to the unimaginable miniaturizing which nanotechnology has led to, it's possible to reduce such systems to extremely small size, and, in the long run, perhaps to have them be placed in brain tissue by nanorobots. We cannot imagine the kinds of hybrids which are about start populating our world. Connecting the human brain with computers and (tele)robots is something which is now about to leave the experimental stage (De Mul, 2014). It is a matter here of cyborgs which go further than having an artificial hip being placed in a living body, as an example of a technical application. We have here a piece of technical apparatus being planted in the human brain, and thereby we see the potential reality of a hybridizing of a technical application and the human mind. It is possible for the one to influence the other in a mutual relationship.

People are saying that nanotechnology, in combination with biology, ICT, and cognitive science, is facing unknown possibilities of crossing boundaries between human beings and machines, mind and matter, and death and life (Oomen, 2010). Thereby, the identity—the wholeness or unity—of human beings is at stake in more ways than one.

This development demands ethical reflection; is all of this really legitimate? It seems unavoidable that this means the obliteration of the boundary between human beings and machines. It is clear, to think in this way is an expression of technical thinking gone astray, thinking only within the categories of materialistic, technical applications. Strikingly, the reductionism involved here is not being acknowledged.

That does not nullify the fact that handicapped or ill people—the deaf, the blind, or the lame—are often able to be helped by these new technical applications, and find that they are able to hear, to see, and to walk. This kind of technology can truly enrich life. But it is also possible that these technological applications can, unintendedly, actually impoverish and mutilate human life when human beings are robbed of their responsibility and even become regarded as mere instruments or tools.

This development inspires some to believe that humanity can be improved by these technical applications—as a form of "enhancement." We must consider if this means that normative boundaries have been crossed which ought not to be crossed. The so-called transhumanists— about which more later—applaud this development and connect to it the dawning of a completely new phase in human history. Kevin Warwick (2004), too, believes that the new cyborg technology represents a reality which will change human evolution.

Being human and technology

It is not surprising that the historical development, as sketched above, of the relationship between human beings and technology has gotten attention from philosophers, and that it has sometimes worked as a catalyst for intense reflection.

The dominant cultural stream of thought, originating in the Reformation after the Middle Ages, which welcomed new technical development for the glory of God, gradually has been replaced by Enlightenment thinking; which seeks answers as to the origin, purpose, and future of our human existence in the light of purely human Reason. In the latter tradition, science and technology have themselves become a kind of path of salvation (Mutschler, 1999). In this cultural orientation, a particular view of Nature has emerged. Nature is seen as a great symphony of powers which can be scientifically analyzed, calculated, and technically managed and controlled. From the start, all of Nature is seen as a complex machine, which, after accurate scientific analysis, can be imitated through technology and even improved upon. No wonder that since the arrival of the modern age people speak about the rise of a mechanical worldview.

Currently, we speak of a technical worldview, because technical applications have to do with more than only mechanisms. Thinking which is oriented to this worldview sees everything through the glasses of (possible) technical management and control. Within that framework, the issue as to the relation between human beings and technical applications is approached.

We can trace the beginning of such reflection to the 17th century. The father of modern Western philosophy—René Descartes—sees humanity as machines, although he still allows room for the human mind in those machines. Later philosophers criticize the idea of the existence of such a human mind as speculation. With de La Mettrie in the 18th century, humankind, with all of our abilities, is regarded as purely *L'homme Machine* ("Man a Machine). With the rise of computer science and computer technology, human beings are by many conceived of as information processing systems. Coming from the dominant technical view of reality, this conclusion is to be expected. Conversely, more and more human characteristics are being attributed to computers. The technical view of being human fosters the idea that human beings and the new technical applications will easily be joined together or result in a true fusion, the cyborg. The differences between human beings and technology will thereby disappear.

This leads the American philosopher Daniel Dennett to say recently, "A human being and a robot are not so different from each other."[1] Dennett is of the opinion that there is no substantial difference between a human being and a robot. According to him, all the typically human characteristics, such as consciousness, are, in principle, possible in robots as well. Even more important, Dennett maintains that human beings themselves are not more than "moist robots"—not made of silicon and steel, but of flesh and blood. Dennett sees consciousness as the result of physical, material processes. "But don't you think that we experience ourselves as a single person, Mr. Dennett?" the journalist of the newspaper *Trouw* in the interview asked. Dennett responds, "That is so. But that is only a story. We are continually constructing a single story from the whole fragmented world around us, about the world and about ourselves. The story of Mr. Dennett, in my case. Dennett is the devil, someone warned one time. In his case, it's not so that the emperor has no clothes, but that the clothes have no emperor!" Dennett grins, "Actually, that's a good statement!"

This view of Dennett's must be seen as the summit of the long Cartesian tradition in which philosophers have thought about the relation between human beings and technology. Let's now look briefly at a number of philosophers in this tradition.

Ernst Kapp

Ernst Kapp is the first real Western philosopher of technology. In his principal work (Kapp, 1877), his main thesis is that we ought to develop genuine insight into the significance of tools as the development of human beings into becoming conscious of themselves. This consciousness of oneself is then to be understood as coming to learn one's own bodily possibilities. In technical applications—making tools—human beings are unconsciously transferring the form and functions of bodily organs to tools. Through the knowledge of tools, human beings learn about their bodies as mechanisms. Kapp acknowledges that being human is more than being a mechanism; the mind transcends the material. Thus, according to Kapp, in a certain sense, human beings are technical mechanisms with a mind.

To give some examples of what Kapp means, "technical applications as projections of organs" take place when human beings recognize

---

1   Daniel Dennett, 2014, in the Dutch daily *Trouw* of Jan. 12, 2014.

a hammer and a handle as extensions and projections of the fist and the arm. The axe is the projection of the fingernail, the drill the projection of the forefinger with its fingernail, the file and saw the projections of a row of teeth, a pair of tongs a projection of the jaw, etc. This principle also applies to more complex tools and instruments. The camera is seen as a projection of the eye, that is, the camera teaches us how the eye works as mechanism. It is more complicated when the construction of a bridge is to be seen as an unconscious projection of the human skeleton, a railroad network is to be seen as the projection of the human vascular system, and a telegraph network as a projection of the human nervous system.

In our day, we could continue Kapp's line of thinking by saying that the computer is an unconscious projection of the human brain, and that the computer teaches us how the brain works technically. From this example, it is evident that Kapp's philosophy has historical significance. We find him cited and used in current concepts concerning human beings and technology (Verbeek, 2009), though, for many, Kapp's attention given to the "human mind" is no longer relevant.

I do not agree with Kapp here. In my opinion, technology can not ultimately be explained by any projection theory. Technical inventions take place, I believe, in the environment of God's creation, and in relation to His governing, cosmic Law, to which all of creation is subject.

There are also inventions which do not fit into the scheme Kapp has constructed, such as nuclear reactors. However, quite often we do see that technical inventions have been inspired by natural examples. Slowly descending drones, for example, were developed in imitation of the examples of jellyfish floating in the water. And the developers of very agile robotic fighter planes were inspired by the very agile, rolling and tilting behavior of fruit flies![2]

Oswald Spengler

The philosopher of culture, Oswald Spengler, wrote, next to his principal work, the *Untergang des Abendlandes* ("The Decline of the West"), a small book, *Der Mensch und die Technik* (1931) ("Human Beings and Technical Applications"). In this book, Spengler sees human beings as predators, which, in line with Nietzsche's philosophy, are dominated by a heroic will to power. Humans are driven by technical applications. Human beings ought not in the first place to be understood as technical beings, but, in this view, their involvement with technical applications demonstrates their struggle with Nature and their will to survive. The tragedy of this development is that humans are victims of their own technical achievements, and with their technical applications perish.

---

2   The Dutch daily NRC, April 11, 2014, p. 19.

Spengler's concept of technology thus fits into his pessimistic concept of culture. In his view, the process of the decline and fall of our culture will be carried out by technology.

The heroic in Spengler's thinking about technology is getting support from contemporary thinkers. This is so, for example, in the case of Arnold Gehlen (1961), who looks at technology in terms of a conception of human beings as being a "Mangelwesen" (a mixed being). That which humans miss through Nature—in the process of evolution—must be supplied by technology. Spengler's culture pessimism is then mostly replaced by optimism. Later, we will see that modern technical applications are above all seen and appreciated as the continuation of evolution.

Here too, I must demur. In all these cases of modern reflection on technology, the thought of them being possibilities within God's created reality is totally absent. This means that there is no room for the idea of a created reality, stamped by evil, but in which Christ has meaning. Christian ethics, anthropology, and cosmology are, sadly, completely missing.

Norbert Wiener

Original, stimulating views of technology have been present since the beginning of information technology with the advent of computers. One example is the book *God and Golem Inc.* by the father of modern computer, Norbert Wiener, about the relation between God, human beings, and machines (Wiener, 1964).

In Jewish folk religion, the Golem is a *homunculus* (very small humanoid) made from clay by Rabbi Löw of Prague about 1580, which could be brought to life by magic words. According to Wiener, the Golem of our modern age is the machine, and especially the machine which can learn and can reproduce itself. This machine is God's co-worker in the ever continuing process of cultural development.

What Wiener says continues to affect all kinds of contemporary concepts. Wiener is of the opinion that the development of the computer sheds new light on the relation between the Creator, humanity, and technical applications. Wiener looks at God's creation of humanity, and human reproduction, through the spectacles of information technology, in terms of what computers can do. The newest machines can learn and eventually reproduce themselves, and this teaches us, according to Wiener, about the creation of humanity and human reproduction. The mechanism of this process can teach us something about the mechanism of other processes. That God created humanity in His image, and that human beings create new human beings in their image, is clear, according to Wiener, through the technical process of computers. God, humanity, and machines are brought together under one category, through this technical conceptualization. This means that God and humanity, in the light of modern machines, are seen as mechanisms which process information, which send out signals, each in their own way, and thereby are dependent on all the signals received up to that point. Information theory expresses this process in mathematical language. New science and technology thus shed new light on God the Creator and on created beings.

We cannot escape the conclusion, if we follow Wiener's thinking here, of seeing God as an information processing system, expressed by a very complex mathematical formula, and following that, seeing humanity as in essence a mathematical expression taking shape in earthly material. In Wiener's view, it is technical evolution—as the successor to biotic evolution—which teaches us to know God and humanity at the deepest level. Wiener is aware that he is hereby launching ideas which will be much more difficult to accept than Darwin's theories about humanity evolving from apes.

At the deepest level of analysis, operating from a biblical perspective, we can say that Wiener's idea of the image of God has been suggested by technological concepts. Wiener's concept is an example of speculation through scientific technical thinking. And Wiener's idea of human beings? They are information processing systems which lose to machines on all fronts. The consequence is that the danger becomes real that human beings become gadget worshippers—worshippers of machines. In order to avoid this worship of computers, robots, and prostheses and in order to prevent machines from ruling over human beings, Wiener supports—happily, inconsistent with his premises—human beings as *responsible beings*. "As engineering technique becomes more and more able to achieve human purposes, it must become more and more accustomed to formulate human purpose."[3] This is why Wiener argues for a symbiosis between human beings and machines—cyborgs—whereby the machines act "in accordance with human values," and human beings make sure that things take place in this way.

Through his philosophy, Wiener has opened the door wide open to allow scientific-technical thinking to become dominant. He has launched his ideas and recognized problems in relation to the possibilities of the computer, which remain the subject of discussion with respect to the relation between human beings and technology.

In the post-Wiener discussions, people are following in his footsteps. However, contemporary thinkers no longer have any use for a degraded, no longer living God, a god who is simply an information producing system, as Wiener maintained. The current discussion about human beings and technology takes place for the most part within the bounds of a

---

3   Wiener 1964, p. 64.

totally closed, material reality, which, according to many people, can be adequately understood in technical categories. Human beings, as created in the image of God, are, in this way, more and more reduced to a mindless part of a controllable technical reality (Schwaab, 2010).

Wiener's ideas are thus continuing to be a stimulation, up till today, in the discussions about the relation between human beings and machines. Humanity has been degraded to a machine—since Descartes—and the machine is more and more seen as a "human being."

In my opinion, this is a dangerous development. I have recently read, for example, as title of an article in the Dutch daily NRC, "The computer says No."[4] The article describes how a computer is making the decision as to whether an illegal asylum seeker may stay in The Netherlands or not. Here, human responsibility has been transferred to computers, and human rights are being sacrificed to the conclusions of technical models. It thus remains for us as an imperative to make sure that technical applications remain subject to human responsibility. This must be clear as we reflect about future technical applications.

Karl Steinbuch

Someone who has even more grandiose expectations regarding how computers and robots are going to be superior to human beings is the German philosopher and engineer, Karl Steinbuch (1965). With the assistance of information theory and computer science, a completely rational analysis of the human mind's capabilities is possible, according to Steinbuch. The functions of the mind include registering, processing, preserving, and reproducing and communicating information. Hereby, Steinbuch is affirming *in principle* the identity of human beings as information processing machines. He downplays the fact that the organic material of the nervous system and the brain differs from the material out of which computers are built. At the moment, this difference of material is the reason why humans are superior to machines, but in the future that will change, in Steinbuch's opinion.

The reason that Steinbuch affirms the identity of human beings and machines, in principle, is that, in his view, we do not come upon an unexplainable "remnant" in human beings when we analyze the functions of the human mind. For him, there is no "secret" of the human mind. In his view, the question regarding the connection between matter and mind is meaningless. Psychological concepts such as affection, fear, anxiety, etc., are justified, in his view, because of the complex structure of human beings, but eventually they can be reduced to physical concepts. Because collective, describable characteristics are able to be identified for humans

---

4   NRC April 20, 2013.

and machines, it is justified, in his view, to speak of the same functions in the case of machines as with human beings, such as thinking, intelligence, learning, reflection, foresight, creativity, the ability to write poetry, consciousness, etc. When he wrote his book (1965), he admitted that not all physical relationships in the brain were known, but he was convinced that continuing research would eventually make that possible. A machine which is not inferior in structure and complexity to the human nervous system would automatically entail that it has consciousness.

According to Steinbuch, when future machines have the possibility of free communication and are able to learn, the possibility exists that machines will surpass humans in evolution. Human "freedom" is thus for Steinbuch only an *apparent* problem. Steinbuch believes that many problems still have to be overcome, because human beings themselves resist this brilliant development by their inadequate ideologies, prejudices, and dogmas, particularly those who say that they are dependent on a transcendent reality, and a "Hinterwelt" ("a world behind appearances"). "Sondern der Gott der Hinterwelt ist tot." ("But the God of the world behind appearances is dead") (1968, p. 50).

However, here we see how Steinbuch is forgetful of his own presuppositions, rooted in a technical ideology. He is a shining example of a technical "believer"; he believes in the power of modern technology to move mountains. And there are many more like him among the thinkers reflecting about human beings and machines. In the current discussions about human beings and machines in the broadest sense, concepts derived from Steinbuch are being elaborated upon. A scientific, technical concept of humanity, and its overestimation—to say nothing of an absolutizing—of scientific-technical thinking, is very evident as the background of this trend.

CONTEMPORARY PHILOSOPHERS OF TECHNOLOGY

The philosophers we have been looking at have had in common a materialistic view of reality, through their predominantly technical thinking, whereby they see human beings as machines and have great expectations of future computers. The philosophers of this moment in time are continuing this pattern, and are even strengthening certain ideas connected with it, and hereby are particularly stimulated by the philosopher Nietzsche with his "will to power" and his idea of the "*Übermensch*" ("superman").

They also share an attitude of not particularly being concerned about the existence of evil. This, while evil has played such a prominently role in human history. Many are evidently convinced that technical applications can overcome evil as such. This, while history ought to teach us to be aware of the possibility of *tyranny* in the use of technology; just look at such things as the degradation of the environment, the problems surrounding climate change, nuclear weapons, and devastating natural disasters, all caused by technical applications fatally failing.

We now turn to a number of our contemporary philosophers in the rest of this chapter; Sloterdijk, the so-called transhumanists, and the Dutch thinkers De Mul and Verbeek.

Peter Sloterdijk

In 1999, the German philosopher Peter Sloterdijk caused a great commotion when he published his book, *Rules for the Human Zoo*. In his opinion, many people follow the tradition that human beings, as descendants of animals, have to be tamed in many ways. But, he asks, "Who or what is able to tame human beings, if humanism, as a school intending to do this taming, has failed?" Sloterdijk means with the term humanism the humanism of the predominant philosophical stream, which accepts ethics as setting limits to technical applications. According to Sloterdijk, this philosophy is dead and must be buried, because it has no answers to the new developments in science and technology.

In his view, the developments in the area of genetics and technical applications that modify human life and the technology of management and control—and renewal—of human beings, demand another approach. Sloterdijk sees no longer the formation and education ("*Bildung*") of human beings in the tradition of the Enlightenment as leading to the improvement of humanity, but is rather looking toward a change of human beings by means of technology. Hereby he goes back to the idea of the "*Übermensch*," as Nietzsche argued for more than a century ago. "It is not our task to tame human beings, but to breed them." The criteria which ought to be used for this will be, according to Sloterdijk, defined by a specialized elite composed of philosophers. By making this suggestion, Sloterdijk is following the ancient Greek philosopher Plato, who sighed once, "Wouldn't it be so much better if all philosophers were kings, and all kings philosophers?" Sloterdijk believes that without an idealized image of the philosopher king, caring for human beings by other human beings will be an idle and useless, though perhaps passionate, endeavor. With philosophers, however, "the memory of a heavenly vision of the very best is most present and active."

Sloterdijk wants to improve human beings through modern technology. He wants to eliminate, once and for all, all appeals to God or to chance. Humanity must exercise its own power in action. "Humanity makes humanity its Higher Power." That was what Nietzsche, too, had in mind in his day. According to Sloterdijk, our day and age is a period in which *political* decisions are called for regarding the future of our race. "Will humanity be able to make the change from birth fatalism to birth-by-choice and to prenatal selection?" In any case, philosophers will have to formulate a kind of ethical codex for technical ways to improve humanity. And "royal anthropological technical applications" must begin by removing "unfit natures" before it will be possible to improve the breeding of human beings. Sloterdijk is a new Nietzsche, as the French philosopher Bruno Latour calls him by way of welcome. The ideas of Nietzsche, the philosopher of God-is-dead, the philosopher of "lawlessness," the philosopher of the "*Unwertung aller Werte*" ("the transvaluation of all values"), are now dressed up in the clothes of the promotion of new genetic techniques and anthropological technology.

Sloterdijk is a super-humanist! With Sloterdijk we see a radicalizing of humanism which distances itself more and more from Christianity. A storm of protest and repugnance erupted in the press in reaction to Sloterdijk's essay. The criticism was that Sloterdijk, with his *Rules for the Human Zoo*, had not learned the lesson of Hitler's Germany, with its practice of eugenics, whereby millions of Jews were murdered. Furthermore, with his idea of letting an elite establish the rules for the "human zoo," Sloterdijk was clearly going in an anti-democratic direction, and waking up the ghosts of a dictatorial, unfree past. The critics rose up in support of freedom, of conscience, of democracy, and of a decent ethics (Habermas, 2003). They saw that, with Sloterdijk, human values are put under pressure or even disappear. Humanity itself becomes an object of complete technical management and control through such anthropological technology. It is a concept which goes even further than fusing human beings and technology.

Sloterdijk has initiated the start of a vital discussion concerning fundamental, burning issues of ethics and technology. Whereas till now, under the influence of Christianity and classical humanism, it was assumed that ethics ought to define the direction of technology, with Sloterdijk and his followers technology itself becomes the *norm* for ethics. As far as that, Sloterdijk is a thinker who clearly overestimates the technical possibilities. His agreement with thinkers who expect to see cybernetic machines with consciousness is in line with his presuppositions. For him, human beings are machines which can be improved upon.

Transhumanists

The thinking of the so-called transhumanists goes even farther in this direction. They maintain that human beings have landed in a post-Darwinian age in which humanity is now the captain of the evolutionary ship. Transhumanists subscribe, in general, to the standpoints of classical humanism, but aim at exploring the farthest limits and even crossing them.

Inspired by Arnold Gehlen (1961), transhumanists support expanding limits and supplementing human shortcomings. They argue forcefully that human beings will have to improve themselves physically, or, analogous to the language used for computers and software, *upgrade* themselves with techniques like nanotechnology and genetic modification and radical integration of computer technology into the human body. Moreover, transhumanists continue to dream of and strive for new technical applications which will defeat *all* diseases, feed the world, and conquer the solar system. Utopia is in sight!

According to Diamendis (2012), nanotechnology will provide for abundant water and raw materials and energy for all the inhabitants of the earth and thereby guarantee their future. The images which we know from science fiction films (Cusveller, et al., 2003) are for transhumanists a genuine reality. Perhaps one day (Kurzweil, 2013) the content of our brains will be downloaded and become immortal. Or immortality will be achieved by means of nanotechnology and through upgrading our brains with computer software in order to conquer the aging process (Drexler, 2013). Transhumanists believe that, through *uploading*, it will be possible in the (near) future to capture human consciousness—the human "mind"—by means of software. Hereby it would be possible to transfer this *software* to another, "better" substrate, such as a very advanced computer which can duplicate all the functions of the brain. This is to *upload* human beings into a virtual world, just as Internet users *upload a homepage* to the web.

The main theme of the transhumanists is thus that humanity is going to be fused with its technical applications. Shortly we will all be constantly connected to the Internet and with each other via technical expansions of our nervous system. For example, this will be possible through microchips implanted in the brain. Eventually, we will be able to enlarge our intelligence to superhuman proportions, and, with the new possibilities given to us, in the long run we will rule over the whole universe and reconstruct it according to our specifications. In short, then we will not be able to be distinguished from what were called gods in the past (Chorost, 2011). This goes together with the idea of colonizing space by these kinds of "*Übermenschen*."

Dutch thinkers

We now look at the ideas of two Dutch thinkers related to the subject we're discussing; Jos de Mul, professor of philosophical anthropology at Erasmus University in Rotterdam, and Peter-Paul Verbeek of the University of Twente. Both of these philosophical anthropologists are on top of the latest developments in technology and discuss them in an engaging and provocative way.

Jos de Mul

Whoever wants to know what is happening in the area of the most recent ICT philosophy should listen to Jos de Mul (2002, 2014). In his publications he describes all kinds of technical developments in a playful way. His assumption is that with the rise of ICT a new phase of the *evolution of humanity* has begun. In line with this, on the last page of his book of 2002 he comes to the following conclusion; "However the development of and cooperation between new disciplines such as nanotechnology, biotechnology, artificial intelligence, and artificial life may still be surrounded by discourse with a human aura... it seems that their agenda is eventually, intended or not, aimed at the creation of a post-human form of life" (De Mul, 2002). As a result, he pens his own *Cyberspace Odyssey* seeing technical applications as an extension of evolution. He thereby elevates *technical* thought to a preeminent position seeing humanity as traveling the road toward becoming the *Übermensch*.

This *Übermensch* will be continually technically connected with the world-brain, as an amalgamation of all the knowledge in the world. From our current vantage point, the *Übermensch* will be unpredictable due to his unfathomable characteristics.

De Mul further elaborates on these thoughts in his recently published book, *Kunstmatig van Nature* ("Artificial by Nature") (2014). In the succeeding chapters about future information technology, nanotechnology, and their mutual relationship, he shows how *homo sapiens 2.0*—that is, contemporary humanity—will make room for *homo sapiens 3.0*. The latter is an android robot and symbolizes a new phase of human evolution through new technology. With the new technology, human beings can be improved upon, changed, and finally replaced. Technically complex robots will be the new humans. This new humanity will be by nature artificial, just as the old humanity, for that matter. What has been transported inwardly as part of being human will evidently be able to emerge into the open.

Again it is more than obvious that De Mul's thinking is *technical* thinking. Thus he does not say—which would be historically defensible— that human beings have always been technically active. No, according to De Mul, the *structure* of human beings is technical in character, and can therefore be continued in a new evolutionary phase. Thereby he links up, in his conceptions, with thinkers who expect that *homo sapiens 3.0*—with post-human intelligence—will choose to leave the planet earth for further development. He sees the possibility that in the future robots will be built which, thanks to artificial brains, will have a form of consciousness and feeling, and which will also develop a free will (2014, p. 168).

The books I have mentioned are written in a gripping way. With great erudition concerning technical developments, De Mul permits himself— as he himself describes it—to pursue a *speculative anthropology*. In this way he opens the door to his own unbridled imagination, leaving a sober analysis of the facts in the shadows. Hereby his books convey a vision which is more surreal than real. The introduction to his second book, in particular, betrays this tendency. Here De Mul is just writing science fiction. He suggests that a human being will be able to marry a human robot, that such a robot could be a delegate to a House of Representatives, and that the rights of robots will be subject to legislation.

It is striking that when he lets his fantasy run wild, he mentions that it is Christians, above all, who oppose his ideas. Although he allows room for the mystery of human consciousness and for the human mind, he does not put any limits on his technical thinking about the future, and puts no limits on the technical possibilities themselves. This is so because he reduces the fullness of reality to an abstract, technical, scientific, material reality and dreams on in it.

De Mul's conclusions elicit protest particularly from Christians, it is true. It is natural for Christians, who accept reality as one created by God—a reality mutilated by humanity but again given hope by God in Christ—to affirm as central humanity's responsibility for technology. This does not mean that they reject all the magnificent technical applications there are, but they continually emphasize that such technology must be made subservient to humanity and society as a whole as a form of service. This normative approach is completely missing in De Mul's writings, although he does speak of possible risks and uncertainties. At the same time, he sees the new technology as something to be taken for granted in the process of humanity's evolution to a higher level.

But what is to prevent *evil* nestling malignantly in this whole process, so that we are confronted, for example, with a kind of Frankenstein monster in the robot of the future? This can certainly be the case if we do not normatively evaluate and, when necessary, put limits on the new technology. We can already see shocking examples of the extremely dangerous, even apocalyptic, effects of present technological applications; for example, in the case of nuclear reactors which are vulnerable to natural catastrophes (Fukushima).

It is true, De Mul himself is not merely optimistic about the results of the technical evolution he foresees. He docs believe that human beings may be the first forms of life which have been destined to create their own evolutionary descendants—superior robots—which are more able than they are to survive in the digital *struggle for life* (2002, p. 320). However, in his view, this *Übermensch* could just as well be an evil threat to, as much as a fulfillment of contemporary humanity. For the birth of a new sort of being is more than just the end of the old sort. This scenario demonstrates both the grandeur and the tragedy of humanity. In the worst case, possible evolutionary descendants may wind up exterminating the human race (2014).

Peter-Paul Verbeek

Verbeek is more moderate in his approach, adopting a different tone to his writing than we hear in De Mul's works (2009, 2014). He emphasizes especially how a new ethical framework is needed with relation to the new technology. For him, modern human beings are *technically mediated beings*. In the classical philosophical anthropology, the human body functioned as a very natural boundary between human beings and technical applications. Now this boundary has evidently become remarkably less clear in the light of the newest anthropological technology. These technical applications do not project bodily functions, nor do they supplement them, but rather they *fuse* with them, according to Verbeek, into a new body.

For Verbeek, the ethics of technology creates room to interact with the new technical possibilities in a responsible way. Verbeek believes that the old ethical approach to good and evil is no longer appropriate to evaluate the new anthropological technology. Instead of maintaining and guarding the boundary between human beings and technology, ethics must now seek to find a *responsible* merging of both. Technology and morality ought to move forward together (Verbeek, 2014).

Hereby he seems to be advocating a more pragmatic approach of a *supervisory ethics*, as he calls it. That sounds very technical and to me suggests too much of an accommodation to the new technology, or perhaps a streamlining of the relationship between human beings and technical applications. Verbeek says that earlier we shifted—through the Enlightenment—the source of ethics from God to human beings. Now is the time for ethics to shift away from *human beings* to *technical things*. "Technology mediates our morality," he says (Verbeek, 2014, p. 14). Technical things belong to the moral community, in the sense that they, together with us, give form to morality. For Verbeek, technical applications are morally charged (Verbeek, 2014, p. 54).

Although Verbeek does not speak of moral technical agents, for him technical applications are nevertheless moral mediators. Evidently he wants to say that we ought to see that the new technology itself has an influence on ethics.

To me, this latter point seems valid. It is certainly true that new technology demands that an effort be made to give shape to norms on the basis of existing values. However, Verbeek does not speak here about a changing relationship between existing values and changing norms, but lets the new technology *prescribe a new morality*, as it were. That's why he ascribes to technical agents a hybrid intentionality and a hybrid freedom. Verbeek has already created a standard hybrid out of present day and future human beings, half human, half technical applications (Verbeek, 2014, p. 73). That is, human beings and technical agents work together. But isn't this in conflict with the fullness of reality as we experience it? Things like rocks and trees do not have an "intentional freedom." Verbeek commits himself to a scientific-technical way of thinking, an abstract way of thinking, which—detached from the fullness of reality—determines his concepts. If we follow him here, the result is that we subject ourselves to the power of technology, and that we lose our own responsibility for them.[5]

In summary, Verbeek's concept can be called a "cooperation with the technical powers." Verbeek hardly expresses any criticism of technology, indeed he gives the impression that he has no problem fully adapting himself to any and every technical development. He thereby confirms without criticism the legitimacy of technical thinking and the technical powers (2014, p. 99). In his ethics of technology, he's concerned with "techniques for interacting with technology" (Verbeek, 2014, p. 108). However, this is to change an *ethics of technology* into a *technical ethics*. "Ethics ought to be much more aware of the ways in which it itself is a product of technology" (Verbeek, 2014, p. 167). Indeed, in my opinion, it is certainly true that an ethics of technical applications is legitimate; an ethics concerned with technical applications is also concerned with the relation between human beings and technology. But Verbeek's *technical ethics* view is a *post-human ethics* which, following the French philosopher Latour's concepts, speaks *anthropomorphically* about technical things (p. 57) as a "morality of things" (p. 63). In Verbeek's writings you can also find him speaking of technological responsibility shared by human beings and technical applications themselves (Verbeek, 2014, p. 124).

Verbeek speaks in this connection about us choosing a "good life." But he is not particularly clear about what direction we should be going. He does promise us that he will elaborate a normative framework in the future. It appears that his ethics of technology is meant to take account of the technical developments under the leadership of Western neo-liberal capitalism (Schuurman, 2004, 2014). In any case, the direction he is choosing does not explicitly contradict materialistic views and the dominant scientific-technical thinking present in the way human beings are carrying out technical management and control. This shows that fundamental values of human beings are being put at risk or can be transgressed. Furthermore, it is striking that the fundamental distinction is not made between a technical influence on a human organ or function and structural, technical influence on the totality of a human being, whereby being human itself is fundamentally changed. In any case, it seems here that an ethics of a factual "fusion" of complete human beings with technology would threaten the existence of humanity.

---

5   See also Carel Peeters in the Dutch weekly Vrij Nederland, May 19, 2014.

A critical evaluation

In my opinion, Verbeek limits himself to the *ethics* of technology,and gives too little attention to the *cultural philosophy* of technology. The other thinkers we have looked at have one thing in common; they expect great things of the new technology. In general they do not have much concern about possible serious problems.

Their expectations of "machines which think," the erasing of the boundaries between human beings and technical applications, and—for the transhumanists—the expectation that the time is coming of the evolution of humanity through the new technology, or that humanity can be improved through the new technology, attest to a certain naiveté or superficiality.

The reasons for this can be traced to a one-dimensional, materialistic view of reality, and to a presumptuous, technical style of thinking. This technical thinking takes place at an abstract level of academic science and at the level of technology thoroughly influenced by such science. This is a *scientific-technological* style of thinking. This kind of thinking about modern technology is, in fact, totally removed from fullness of reality, and is leading many people onto a path which is fundamentally going nowhere. This approach deeply colors the current debate about *human enhancement* and the thinking of transhumanists like Kurzweil and Drexler. But De Mul, as well, is more surreal than real, estranged from reality.

The thinkers we have looked at claim to be speaking about reality in its fullness, but have reduced it to what can be scientifically theorized about or carried out in new technological applications. The consequence of this is that reality in its fullness is transported to the level of scientific abstraction, and thereby is distorted or mutilated.

Can computers think?

The issue of whether computers can think or not is illustrative here. The mathematical philosopher Alan Turing (1960) is one of the first who believed that we can affirm that computers can think. He developed a so-called "imitation game." If on the basis of answers given to questions addressed to a human being and to a computer, both invisible, it can not be decided whether the human being or the computer gave the answer, we would be able to conclude that computers can think, and thus are intelligent. Since Turing's presentation, this conception has dominated the discussion of this issue.

There is something strange going on in this discussion. This is because Turing's argument, on closer examination, is not at all convincing. He demonstrates through his imaginary test what already has been assumed to be the case. He compares human beings and computers, seeing them as two information processing systems. In his hypothetical test, human beings are completely conditioned by the possibilities of an information processing machine. That's why in the test the human beings and the computers may not be seen and may only receive questions to which they can answer "Yes" oo "No." We see here that Turing conceives of the thinking of human beings as being totally formalized, that is, able to be established by a set of rules. As result, this thinking can be imitated and improved. We can speak of "artificial intelligence" if we are careful to keep emphasizing the "artificial" character of this intelligence.

Why do people want to affirm that computers can think? Because it fits into the mainstream of Cartesian philosophy. Since the Modern Age, thinkers have been expecting that technical applications will be able to free us from our human shortcomings and limitations. Within the framework of abstract, scientific-technical thinking, anthropomorphic language concerning human beings—humans think—has been applied to the computer without any reservation. Computers can think!

However, in reality what is happening is that people are looking from this vantage point at humanity "*computer-morphically.*" Human beings are being looked at through the spectacles of the possibilities of the computer. Herein there is a reduction of thinking and thus a reduction of what it means to be human as well. In this process, people start by ignoring the big distinction between human beings and computers. And in making the comparison between human beings and computers, people regard the computer as an independent whole, while, in fact, the computer in itself does not exist. For it is clear that every computer is made and programmed by human beings. Because computers often work in apparent independence, this is easily forgotten. And because succeeding generations of computers are getting smaller and smaller in size, working faster, having more storage capacity, and shortly will be integrated into neural networks, and at the same time will be taking more tasks over from human beings in robots, they give the appearance of having intelligence. They make comparatively few mistakes (once bugs are eliminated) and are becoming more "intelligent" in carrying out functions. Furthermore, they are tireless and surprise us with promising results.

It is true. Computers are an amazing new technological invention. Indeed, many human activities, after analysis, can be objectified, and, via software be programmed in computers can thus be taken over by computers and strengthened. Integration with information gathered through artificial sensors provides us with robots able to use artificial limbs. These robots can learn from their own "experiences" and improve themselves. And these developments continue. I imagine that robots of the future will be able to do far more than many of us deem possible. Nevertheless, I would rather not speak of "thinking" computers or "anthropomorphic" robots. The computer ought to be seen as always remaining in service to humanity, a product of human ingenuity and skill, and a human creation, whether as a technical tool, or a tool for giving guidance or helping to think through problems. Calling this "tool activity" helps us avoid conceiving of computers as being truly independent, and prevents them ever from being placed over against humanity. Making them truly independent is actually impossible, even in the case of the robots which we have sent to Mars. They can do a lot that human beings cannot do—that is even their purpose, and that's why we use them—but they are meant to always be controlled by human beings—even at a distance.

But in the summer of 1997, wasn't chess grandmaster Kasparov beaten by the super chess computer *Deep Blue*? Isn't that proof that such a chess computer can think (play chess)? Don't fool yourself. It was not the chess computer which was conscious of the victory, and was happy to win. It was the many programmers involved who *were*! Because computers operate apparently so independently, we can easily come to wrong conclusions.

Instead of saying that the computer could play chess better than the best human chess player, you actually ought to say that the programmers of *Deep Blue* knew how to construct chess playing software which could beat Kasparov. Talking of thinking computers and chess playing computers keeps the myth about unimaginable technical progress alive. If a computer could say about itself, "I'm thinking," or "I've won a chess match," without having been programmed to use such sentences, we're inclined to laugh. For everyone is aware how computers lack true consciousness.

What computers and robots can't do

In the meantime, the achievements of computers and robots are very impressive. New applications are constantly being developed. We see this in different areas, such as making diagnoses in hospitals, keeping track of stocks in warehouses and stores, and even preparing and carrying out social decisions via rules and protocols (Brynjolfsson and Macfee, 2014). There are good reasons for seeing computers and robots as belonging to the 50 machines which have drastically altered human history, with the expectation that they will continue to do so (Chaline, 2012).

This technical development has many social consequences. Certain human tasks are being taken over by machines. Gathering a large amount of data—a time-consuming activity for human beings—is advancing, and human beings who work according to old patterns and needs are becoming "redundant," thus unemployed. This, unless they choose (at an early date) creative tasks which are involved at the beginning of a technical development and which cannot be taken over by new technology. In the end, all technical discoveries and inventions come from human beings. We see, for example, that the social media computers play a big role in the management and maintenance of the systems, but that the possibilities involved in Facebook, Twitter, and Wikipedia are clearly the conscious creations of human beings. And it is self-evident that computers and robots can never independently answer questions about meaning and purpose in life.

What is creativity? Creativity is nourished by curiosity and is characterized by persistence and unexpected new views of possibilities in new combinations. Further characteristics are a flexible style of thinking, playfulness, youth, a questioning mind, the capacity to look at customary things in a different way, daring to depart from traditional solutions, a high level of intuition, a measure of introversion, and a great intellectual capacity, coupled with bravery. Via "brainstorming" and dealing with criticism, creativity can be enlarged and released to make creations. Thus it is possible for creativity to be developed and stimulated by an education which does not merely travel the trodden paths. This kind of developed creativity is the basic prerequisite for taking new paths, passing boundaries, and creating new possibilities. Creativity chooses a *strategy* in order to achieve something. *That* is a task the computer or the robot can never take over. Everything which has to do with subordinate tactics within a strategy can often be very well carried out by a machine, in service to its human makers. That is the big difference.

Furthermore, creativity is not always beneficial. It is clear that creativity can seek an alliance with and even be possessed by evil. Just look at human history. The malevolent inventiveness of creative human evil, evident in violent conflicts and warfare, in destructive human relationships, in criminal activity, in economic exploitation, in human trafficking, in sadism, has certainly not been diminished in our cyber age. Creativity is not, in isolation, a blessing.

What is typically human with or without striking, innovative creativity? Every human being has access—by insight or intuition—to the essence, the substance of things. The computer can never do that. Having insight into the essence of something has been given to all human beings so that everyone can say "I"—from a spiritual center. In the terms of reformational philosophy; the computer acts as a *subject* in the spatial and physical-chemical functional spheres, but as an *object* in the other spheres of reality where human beings function as the subjects, such as choosing something new, acting in the economic sphere, acting in the esthetic sphere, seeking justice, expressing care and mercy, showing love, trusting, and believing. Machines will never be able to have vision, wisdom, love, patience, and empathy the way human beings do. In all these areas, human beings are superior to the most modern technology, and as such, we continue to bear responsibility for all technology.

Looking at things in this way, the revolutionary technical developments around us can certainly lead to a more "human society" in the sense of fitting God's intent for human life together. The fact that the reverse happens regularly is often because people let themselves be "encapsulated" by technical applications and don't know how to escape a reductionist technical way of thinking. Technical thinking can be enriching if it is not serving evil and is not reducing reality to that which is merely material, but is allowing room for that which is spiritual and for the unity of all created things in their divine origin. People who still have their "heart" in their "heads" and in their "hands" will never be superfluous in this technical age! It is further true that creativity will continue to be needed in order to make space for meaningful endeavors available to many (Dessauer, 1956, Schuurman, 2004, 2009).

A Christian, philosophical view

The expectations of contemporary technology thinkers concerning so-called anthropological technology are based, as we have seen, on an extremely reductionist and superficial image of humanity. To derive everything which is human from material, physical structures and to interpret these in a technical way is to do injustice to the complex structure of human beings and thereby to lose the means of resisting possible threats to human existence by technical applications.

Christian philosophy can be mobilized as a weapon against wrong expectations and overestimations of the possibilities of technology. In a structural analysis of human beings, it is an essential fact that human beings are much more than physical structures. Human beings are a woven tapestry composed of a physical structure (atoms and molecules), a biotic structure (cells), a psychological structure (the nervous system), and a human, normative act structure; the *I*, which wills, thinks, and believes, and which refuses to be captured by theoretical analysis (Ouweneel, 1986). This "I" is connected to the image of God in the Bible. This image of God qualifies the center of human existence, and from that center determines all responsible activities of human beings; "...for from it (the heart) flow the springs of life" (Proverbs 4:23b, ESV).

In order to see this, we have to distance ourselves from the dominant scientific-technical way of thinking and the view of humanity and reality which flows from it. How can humanity still expect a human future, if, from the start, human beings are seen as being machines? This reductionist way of thinking implies a reduction of human activity as it truly is, a being free-in-responsibility. The consequence is that a meaningful perspective is excluded a priori.

This does not mean, however, that technical applications cannot be woven together into human activity. We see this in all examples of physical prostheses. The complex structure of human beings is, in these cases, woven together with a technical apparatus. This pattern can certainly be extended via modern technical applications, for example, at the level of bodily organs and functions. But the implementation of such technical structures—and here we see boundaries coming into the picture—must be careful to respect the inviolability of the human body and the protection of the personal, spiritual identity of every human being. Protecting the wholeness of human beings does not allow for humans being woven together with technology at that level.

So, for example, it would be legitimate to make use of genetic modification techniques to modify bone marrow cells in order to foster recovery from leukemia. But the same technique used to change someone's DNA structure would be illegitimate. The latter kind of a "weaving together" would be taking place, for example, when by means of the germ line the genetic structure of (future) human beings would be manipulated, without realizing the possible negative consequences.

Because many people are thinking about robots and cyborgs without taking into account human freedom, humanity is being subjected to a priori technical management and control style of thinking. It is precisely in order to safeguard human beings from presumptuous, reckless thinking, and to do justice to human responsibility, that human freedom *must* be presupposed in doing science. That is to truly champion humanity.

Many thinkers concerned with technology do not really value a free and responsible humanity. In theory, a humanity without freedom or responsibility is already the victim of the machine. This is the result of having an ideal of limitless scientific-technical management and control. This ideal, in a dialectical way, elicits opposition in the form of antagonism to the threatening character of a technological world by those asserting the need for human freedom. When freedom is not acknowledged as the boundary of thinking, we see that a cry for freedom breaks forth and demands to be heard.

Whereas De Mul in his books is very clear about the basic identity between human beings and robots, in TV discussions you don't hear him making the same points with such clarity because the interviewer tends to assert the need for human freedom—and thus the uniqueness of human beings—in vivid terms. De Mul reluctantly concedes the point, and becomes unsure about what he wants to say.

Although modern technology is coming too close for comfort for many people (Van Est, 2014), when we think about it we must continue to affirm that humans are *free and responsible beings*. It is clear that human beings always think in terms of freedom, evaluate freely, and make decisions in freedom. In the case of human beings possibly being woven together with technology, it is imperative that humans must be able to continue to think and act in this way. Such technology may not be allowed to make this impossible. While this human freedom is conditioned by all kinds of factors, it is also normed, and that is precisely the grounds of its possibility; *human beings are by definition free and responsible beings.*

Technological applications are developed by human beings using their freedom, and thus tools which are made to serve humanity. These powerful tools are things by which human cultural products are objectified and strengthened and afterwards used meaningfully by humankind. We see around us how machines and robots can do more and more on the basis of increasing scientific-technical insight. That's a beautiful thing! When through this process, human responsibility is not put under undue pressure, but retains its leading role. This is something valuable and makes progress possible. This also applies to human bodies being woven together with technological applications. In this case, we must always take into account normative boundaries which may not be crossed, such as protecting people's bodily integrity and personal, spiritual identity. If this kind of protection is present, then we can speak of restoration in relation to anthropological technology (using the term anthropological here to refer to operations affecting human beings) —such as seeing, hearing, and moving—and even sometimes helping human existence to emerge and develop.

If this is not the case, the consequence is the distinct possibility of a serious disturbance of or even a threat to life. Then, technology is not serving life, but undermining it. This can be so, for example, when an implanted biochip in the brain is controlling the person involved and is threatening his or her will. Unfortunately, most of the thinkers who focus on the new technology in this area do not offer any resistance to these developments.

In a Christian, philosophical approach, seeing the intensification of technological applications and the results of them. We will call for not less, but more human responsibility to be exercised. This offers the best resistance to the threat of abuse. Responsible freedom creates the possibility of norming the new technology, seeking to manage it, and using it beneficially. Furthermore, human beings are continually able to achieve something new, technically speaking, on the basis of the mystery of their creativity. In this way, human responsibility does not disappear, but increases and becomes increasingly social in character.

The discussion should moreover be addressing the content of the concept of cyborg, and with the issue as to how technologies incarnate values and norms, and change our experience and ideas. How we give shape to technology ought to be on the agenda. We need an *ethics of responsibility.* Indeed, the new technical applications demand a forging of new norms on the basis of established values. The direction of such a responsibility ethics will be determined by the question of whether the new anthropological technology will enrich human life or not.

The present political landscape

Technology creates an enormous cultural power, and via the ongoing developments it has a great influence on human life. It is therefore not surprising that politics is involved with new forms of technology. For example, there is justifiably a great deal of concern that human privacy is put in danger through participating in cyberspace via computers. More protection is needed. In another example, political concern was called for when the possibilities of genetic modifications became known. That will also be called for now that anthropological technology, as we have seen, is receiving attention.

At the beginning of these developments, there were large ethical differences of opinion concerning genetic modifications. Those differences of opinion still exist, but there is now evidence of progress being made in the political arena here in The Netherlands, in reaching a degree of consensus. Such a consensus is increasingly to be seen—at the European level as well—regarding genetic modifications of agricultural crops. Modification within the same kind of plant—"cis-genesis"—can count on more and more support, while there are, correctly, still many objections to "trans-genesis," because of the many unknown risks. And regarding the application of this technique to humans, for example, to heal diseases, there is increasing agreement supporting treatments at the level of organs, while practically everyone rejects this treatment via the germ line. This is so because such treatments affect the *whole* of the human person, and their results are uncertain.

Regarding stem cells, a similar response is quite common. The idea of using one's own stem cells has not caused any opposition among political parties in The Netherlands. But if such cells are to be taken from human embryos, especially grown for that purpose, then there are fundamental disagreements between the political parties. A way out of this problematic situation might be found if the procedure would involve bringing existing cells of someone's body back to the state of undifferentiated stem cells. The healing of many diseases is then conceivable. Happily, many people, in and out of politics, reject cloning human beings, because in the cloning process the *whole* of the human person is involved. From this it can be concluded that the distinction between techniques which involve the whole person and those which affect only an organ or a function is a fruitful one which remains promising for the future.

With relation to the development of anthropological technology, political parties will have to become more alert. When technical applications are once again aimed at the *whole* human being, or these kinds of techniques threaten to undermine the integrity of the human person, or put the personal, spiritual identity of the person under pressure, alarm bells should start to ring. This should certainly be the case when ideas in the realm of eugenics or other suggestions under the influence of transhumanists are being seriously entertained. Then we must stand up to defend human rights. Alertness is demanded of political parties with relation to these new techniques, and we are called to constant vigilance in this area.

When we address the issue of robots, we must be aware that robots are already not always able to be managed and controlled by human beings due to their complex software, which is not always transparent. This is even more so if robots be allowed to reproduce themselves. It is actually very surprising that political parties have not already put a brake on this development. The 21st century technical applications—genetic modification, nanotechnology, and robotics—are so powerful that all kinds of unforeseen accidents and forms of abuse can occur. Furthermore, politicians are justly concerned about nuclear waste winding up in the hands of terrorist individuals.

This can also happen in the case of the new technology, yet we hear very little about these latter dangers. Many people are now chasing after the dreams of material promises created by new technologies, within the system of world-wide capitalism, with its alluring images of financial success and competitive advantages. Critical voices are heard too little, and the real concerns which we should have are not really shared sufficiently (Joy, 2000). Furthermore, we are faced with new Big Brother scenarios and techniques affecting (mainly unconsciously) our behavior, which demand ethical reflection and public, political, debate.

In short, careful, critical reflection about the new, often promising technology of our time must be aimed at having Creation be creatively unfolded and allowed to flourish, seeking the glory of God the Creator, the well-being of all human beings, and the protection of all living creatures.

Literature

Brynjolfsson, Erik and McAfee, Andrew, (2014), *The Second Machine Age*, Cambridge, Mass.: MIT Press.

Chaline, Eric, (2012), *Fifty Machines that Changed the Course of History*, Hove, England: Quid Publishing.

Chorost, Michael, (2011), *World Wide Mind: The Coming Integration of Humanity, Machines, and the Internet*, New York: Free Press.

Cusveller, Bart, Verkerk, Maarten, De Vries, Marc J., (2007), *De Matrix Code: Science Fiction Als Spiegel van de Technologische Cultuur*, Amsterdam: Buijten & Schipperheijn.

Dessauer, Friedrich, (1956), *Streit um die Technik*, Frankfurt: Knecht

Diamendis, Peter H., and Kotler, Steven, (2012), *Abundance: The Future is Better than You Think*, New York: Free Press.

Drexler, K. Eric, (2013), *Radical Abundance: How A Revolution in Nanotechnology Will Change Civilization*, New York: Public Affairs.

Van Est, Rinie, with Rerimassie, Virgil, Van Keulen, Ira, and Dorren, Gaston, (2014), *Intieme Technologie: de Slag om ons Lichaam en Ggedrag*, Den Haag: Rathenau Institute.

Gehlen, Arnold, (1961), *Anthropologische Ansicht der Technik*, Düsseldorf: Verlag Schilling.

Habermas, Jürgen, (2003), *The Future of Human Nature*, Cambridge: Polity Press.

Haraway, Donna A., (1991), *Cyborg Manifesto*, New York: Routledge.

Joy, Bill, (2000), "Why the Future Doesn't Need Us," in *Wired Magazine*, April, 2000 edition.

Kapp, Ernst, (1877), *Grundlinien einer Philosophie der Technik*, Braunschweig: George Westermann.

Kurzweil, Ray, (2013), *How to Create a Mind*, New York: Penguin.

De Mul, Jos, (2002), *Cyberspace Odyssee*, Kampen: Klement.

De Mul, Jos, (2014), *Kunstmatig van Nature: Onderweg naar Homo Sapiens 3.0*, Rotterdam: Filosofie van de Maand/Lemniscaat.

Mutschler, Hans-Dieter, (1999), Technik als Religionsersatz," in *Scheidewege: Jahresschrift für skeptisches Denken*, Vol. 28, (1998-99), p. 1-22.

Oomen, Palmyre M.F., "Vragen over Mens, Techniek, Natuur en God: Filosofische en Theologische Overwegingen bij Nanotechnologie," in *Nanotechnologie: Betekenis, beloftes en dilemma's*, Nijmegen: Valkhof Pers, p. 143-175.

Ouweneel, Willem J., (1986), *Proeve van een Christelijk-Wijsgerige anthropologie*, Amsterdam: Buijten & Schipperheijn.

Spengler, Oswald, (1931), *Der Mensch und die Technik*, München: Beck.

Schuurman, Egbert, (2009), *Technology and the Future: A Philosophical Challenge*, Grand Rapids: Paideia Press.

Schuurman, Egbert, (2004), *Geloven in Wetenschap en Techniek*, Amsterdam: Buijten & Schipperheijn.

Schuurman, Egbert, (2014), "Transformatie van de Materialistische Maatschappij," in *Radix*, Nr. 1, p. 41-53.

Sloterdijk, Peter, (1999), *Regeln für den Menschenpark: Ein Antwortschreiben zu Heideggers Brief über den Humanismus*, Frankfurt am Main: Suhrkamp.

Steinbuch, Karl, (1965), *Automat und Mensch*, Berlin: Springer Verlag.

Steinbuch, Karl, (1968), *Falsch Programmiert*, Stuttgart: Deutsche Verlagsanstalt.

Verbeek, Peter-Paul, (2009), *De Grens van de Mens: Over Techniek, Ethiek, en de Menselijke Natuur*, Inaugural lecture, University of Twente.

Verbeek, Peter-Paul, (2014), *Op de Vlveugels van Icarus: Hoe Techniek en Moraal met Elkaar Meebewegen*, Rotterdam: Lemniscaat.

Warwick, Kevin, (2004), *Cyborg*, Champaigne, Illinois: University of Illinois Press.

Wiener, Norbert, (1964), *God and Golem, Inc.: A Comment on Certain Points where Cybernetics Impinges on Religion*, Cambridge, Mass.: MIT Press.

Swaab, Dick F., (2010), *We zijn ons brein*, Amsterdam: Atlas Contact.

Turing, Alan M., (1950), "Computing Machinery and Intelligence," in *Mind*, Vol. 49, p. 433-460.

Turkle, Sherry, (2011), *Alone Together: Why We Expect More from Technology and Less from Each Other*, New York: Basic Books.

IJsselstein, Wijnand A., (2013), *Psychology 2.0: Towards a New Science of Mind and Technology*, Inaugural lecture, Eindhoven: Eindhoven University of Technology.

CHAPTER 2

MODERN TECHNOLOGY, THEORY, AND WORLDVIEW[1]

Only a few modern thinkers have plumbed the depths of the ideology of technology. Think of the French cultural philosopher Jacques Ellul and the German-American theologian, Paul Tillich. In their criticism of our "technical society," they have called attention to the distinct spirit of technological development in the West. We see in their analysis a deepening and a widening of a perspective on contemporary culture. They show how, for example, education, the justice system, ethics, and even religion threaten to be taken over by an absolutized technical attitude—which in various places they refer to as "instrumentalism." Furthermore, they make clear how technical fever is the cause of much social and economic misery (Ellul, 1989, Tillich, 1985).

The technical worldview a modern phenomenon

We can speak of a distinctly present-day, *modern technology* in contrast to the older, *classical technology*, in the sense of the classic craftsmanship of the past. Looking at this historically, we see some clear differences between the craftsmanship of the past and modern developments.

We cannot really speak of the classical artisans of the past as having a technical worldview. While many were religious, their religious worldview was not technical in character. Such artisans, such as stone-workers, were skillful and creative, but their productions did not reflect a view of reality defined by technical knowledge alone. Technical craftsmanship is made possible in all ages, with or without a technical worldview, by human beings' innate physical abilities. Such craftsmanship is a fully human endeavor. It is dependent on the tools available, and, in my view, it is always part of a greater cosmic whole, a created world, which in itself is not technical in the normal sense of the word.

---

1   Edited version of a lecture originally given to a gathering of the student organization Civitas Studiosorum in Fundamento Reformato, held in Utrecht, The Netherlands, in 1996.

Technical craftsmanship, in all ages, can have a tremendous, magnetic effect on people. And unfortunately, the results of the power and art exercised in applying such knowledge can even become objects of adoration or worship. We see this vividly illustrated in the biblical account of the tower of Babel in Genesis 11. We discover this adoration today when we look at the reactions to what modern technical engineering has produced around us. Many thinkers, in contrast to Ellul and Tillich, have seen modern technology as conveying an inspiring message, leading to the ultimate "salvation" of humanity itself, a kind of modern Gospel for the nations. This shows a continuity between the modern technical world and the old world of craftsmanship (Genesis 11); in both cases ,human beings adore and worship awe-inspiring objects constructed by human ingenuity.

But a discontinuity is evident too. In our day, modern technical applications have become part and parcel of a total technical worldview. This worldview is a deeply captivating vision for many, even more powerful than what we see happening with the tower of Babel. That is, in our day, in the West and in much of the East, everything tends to be seen through the spectacles of modern technology. Think of the pervasive presence of the Internet. Everything in reality has become the object of scientific-technical domination, that is, invasive human management and control by technical means. The 20th century German philosopher Heidegger said, correctly, that our age is truly the secular and technical *Zeit des Weltbildes* ("the age of worldview"). This was not the case prior to the age of modern technology in the West. Heaven had not yet been "nailed shut" by human thought constructions.

How did this happen? The world of technical engineering, brought into being and stamped at its core by natural science, has distinguishing characteristics which derive from and propagate a particular worldview. Both natural science and the engineering sciences, carried on at universities, share a convincing, apparent universality and logical necessity, describing reality with comprehensive mathematical formulas which define the laws of Nature and show how these laws can be put into practice to radically shape human life. When these characteristics are then concretely transferred to technical applications, this leads to a mainly unconscious but nevertheless comprehensive commitment to a technical worldview. That is, modern technology, with its specific characteristics, becomes an all-consuming model for the entirety of reality.

It is important to observe that this takes place only when technology as a scientific study is applied as widespread technological products and systems. In fact, a particular technical worldview (such as the current dominating one) is not in itself inevitable in human history. There is certainly sufficient room for carrying out responsible technical applications in our world by those holding various, different (religious) commitments and worldviews. However, we should not underestimate the seductive power of the modern technical worldview on all within its reach in the West and in the East.

Historical developments in the West

The roots of the modern technical worldview in the West can be traced back to developments in the western world following the Middle Ages. Western thinkers began more and more to regard humanity, not God, as the center or even the fulcrum of reality. On this standpoint they proclaimed human beings to be the real "Lords and Masters" of existence.

It can be legitimately maintained that today's dominant intellectual, ideological movements in the West are the fruits of the Renaissance and the Enlightenment, culminating in the latter's dictum, *ni Dieu ni maître*!("No to God and to all masters!"). These modern movements with this revolutionary background attempt to achieve hegemony, that is total dominance, over all human activity via science and science-oriented philosophy. Today these movements are continuing to spread out vigorously across the cultural landscape of the entire world, like swarms of locusts. That is the chief reason, in my opinion, why our Western culture, besides being a technical culture, has also become a thoroughly secular, godless one.

It is striking to me that often, when people come to recognize the ideological connection between our man-centered, technical culture and the arid, individualistic secularism around us, in the light of historical developments in the West, they fall strangely silent. They are struck dumb by this discovery. It feels like an irreversible, inevitable development, and they feel paralyzed by it.

The green turn and the dialectic involved

The power of the technical worldview is undeniable in our day. There are many who recognize the dangers of the extreme anthropocentric character of our culture, and look for alternatives. Sometimes, in reaction to making humanity the measure of all things, a "green ecological" approach is taken, seeking to bring back health to the plants, animals, and natural environment of earth.

Many are then caught up in a back and forth, dialectical movement between a rational, scientific-technical view of Nature, describable by mathematical laws, and a non-rational, idealized, romantic view of Nature, the serenity of the Serengeti plain (Cobey, 1990). This dialectical movement, in my opinion, demonstrates the underlying restless, neo-pagan character of our secularized culture, oscillating between allegiances like a tennis ball in play.

Rejecting God

I believe that this dialectical movement arises, at the deepest level, because we human beings fail to listen to God and His revelation. We begin making created things central, in contrary, often contradictory, and thus dialectical ways. It is my deepest conviction that we in the West are actively ignoring the reality that the world is God's magnificent creation. We then blithely deny the depth of evil in our midst and in ourselves, and we refuse to believe that there was a historical Fall into disobedience, with all the dark consequences that entails. What is more, we painfully ignore the sweeping redemption Christ has brought into being by His life, death, and resurrection. And we tragically lack the hope given by a vital expectation of a new creation to come.

The dialectical movement in our culture between opposites (a scientifically analyzed, rational, Nature of impersonal laws, and an idealized, romantic, even pantheistic Nature) is at bottom a kind of malignant parasite, feeding on the reality of God's creation and of Christ's redemptive might in this world, but in the process becoming self-destructive.

Kuyper's blind spot

We return to the matter of modern technology. Modern technical development is accepted by many of us in a much too uncritical way. This was true already for the great orthodox Protestant theologian and politician in The Netherlands, Abraham Kuyper (1837-1920) in 1891. Although he grounded his "architectonic critique" of society in the perspective of eternity, he said very little to relativize the importance of the technical developments of his day. That was to some degree understandable. Beneficial technical advancements were everywhere visible. Who could deny that they manifested enormous potential for eliminating poverty, hunger, and disease, as well as opening up possibilities for a spiritually richer culture? Weren't such developments undeniable signs of the Kingdom of God? Kuyper saw, correctly, that we *do* need technical advancements. But he had little insight into the dangers of turning technical developments into an ideology—the awesome power of the technical worldview. Kuyper wrote in *Pro Rege* of modern technical achievements like the telephone as being even greater miracles then the miracles of Jesus! (Kuyper, 1891, 143 ff.)

Kuyper forgot the biblical warning (the story of the tower of Babel in Genesis 11) that human beings continually attempt to dethrone God Himself by means of their technical prowess in order to make a name for themselves here on earth, as they try to build a culture-without-God, a Babel culture. Strikingly, Kuyper's evaluations of technical advancements were exclusively laudatory and they had the effect of blinding his later followers. It took a long time before modern technology began to be critically analyzed in The Netherlands.

Although the praise of technology in our own day is no longer undiluted, criticism expressed does not go very deep. There is a desperate need for a renewed, in-depth, critical analysis from the perspective of eternity, à la Kuyper, but then, unlike Kuyper, aiming carefully at the technical worldview being propagated around us. Only then can this worldview be broken open and released from its bondage to immanent, purely materialistic thinking.

At the 1991 commemoration of the 1891 First Christian Social Congress, held in The Netherlands, the dangers of modern technology were pointed out, but the background of these dangers was not really examined. This is precisely the specific background which is so illuminating.

Is it the economy, stupid?

As a rule, much social criticism of our day regards technological applications as neutral things which only become dangerous when operating within a malignant economic and social order. It is undeniable that when we look at the problems of the Third World, a critique of exclusive free-market economics is warranted. Causes of world-wide labor problems, environmental pollution, and the destruction of the biosphere can be uncovered by such analysis. However, in my opinion, deeper insight into the problems can be gained only if the criticism of the ideology of free-market economics is vigorously supplemented by a criticism of the ideology of modern technology.

In fact, it can be fairly argued that the current technical worldview,or the ideology of technology—I would call it the spirit of the *absolutizing* of technology—in fact precedes and underlies the ideology of free-market economics (Schuurman, 1985, 9-30; 1989).

Seeing only economic factors as the motor behind political developments in the world is inadequate. Think of nuclear arms technology and the development of space exploration technology. Why have huge, disproportionate investments been made in these areas? Benefits to the economy were not at all primary. Rather, these developments reflected the competition between the reigning world superpowers that seek glory and honor for themselves.

Further, the effects of the technical worldview are broader and deeper than those which can be tied to any economic theory. Many scientific and academic fields, such as the study of law, psychology, and even theology, express themselves often in scientific-technical categories. Using economic categories is regarded as neither scientific nor appropriate. Here too, economics bows to technology.

Given this state of affairs, how is it possible that the modern ideology of technology has received so little attention, up to and including Christian circles, when looking at social problems? Could it be because there are many articulate defenders and opponents of market economics—in the line of Adam Smith and Marx—but not many vocal, prominent defenders or opponents of an ideology of technology (although of course articulate defenders and opponents do exist, as we have mentioned and will explore later)? And could the reason for that be that the ideology of technology actually needs no striking, visible proponents because it is so pervasive?

It seems to me—as I suggested earlier—that technical applications, old and new, have formed a constant, seductive power since the Fall into sin, and that this power has been enormously reinforced in a subtle but powerful way by the influence of modern science on technological development.

Utopia or dystopia?

Starting in the Middle Ages and forcefully increasing in the Renaissance, Western society, more and more imagining itself autonomous, became enthralled with modern technical advancements. Indeed, this vision can be discerned already in Medieval and Renaissance dreams of utopia, inspired by the possibilities of technical applications. We may think of the utopias of Roger Bacon in the 13th century and Francis Bacon in the 16th (*Nova Atlantis!*) (Ihde, 1985). While himself a Christian, Francis Bacon expected that many results of the Fall into sin could be effectively surmounted by future technical achievements—great expectations indeed!

Some thinkers of our day conceive of the entirety of reality as being able to be harnessed by technical means and are aiming for utopia on earth, echoing the two Bacons of the past. Indeed, a humanity without God (unlike the two Bacons mentioned, who were theists) tries with its immanent, scientific-technical power of control and management to be Lord and Master over all of existence. The perspective of eternity has disappeared completely, signaling a deep rupture between the divine world and earthly reality.

The results of this development have become evident. Especially after the Second World War, many ordinary people in the West began to admire the seemingly unlimited technical advances around them. They beheld the evident material success which technology provided. Materialism and consumerism were the results.

In my estimation, there has never been an age such as ours in the West, which is so thoroughly technical, but there has also never been an age which has been so spiritually empty. Redundancy—no jobs available for the technically unskilled, resulting in, for some, homelessness, widespread depression, alcoholism, obesity, a small, luxurious upper class pursuing hedonistic futilities, restlessness and ennui, loneliness, burn-outs, a youth drug culture with drug addiction as the result, families falling apart (40% of marriages ending in divorce), living without any limits, and "killing time" are some of the expressions of a process of a cultural landscape turning into a smoking wasteland. And, ironically and unconvincingly, when polled, the majority say they are happy. This, in my view, is the dystopian, bitter fruit of human pretensions, committed to deifying technical power.

If the godless devour the righteous, God is silent, says Habakkuk (1:13).

Technological hegemony

At the moment the ideology of technology has captured and is holding hostage the very core of our culture. It lays a virtually absolute claim on our society. Due to this development, many sectors of culture are fundamentally determined by those "in charge," those exercising scientific-technical control and management in government (with their extensive bureaucracies), education, industry, and business. This is also true for agriculture, health care, and social work.

Ironically, people often try to solve the problems which technology has caused with more technology! For example, look how information technology, biotechnology, and genetic modification—or manipulation—are automatically presented as self-evident solutions for the problems of the environment, for labor problems, and so on. But are we on the right track?

By now, we ought to have come to the realization that the problems regarding unemployment, occupational disabilities, the pollution of the environment, and seeing labor as merely economically productive labor—forgetting other forms of human labor such as communications, artistic creativity, and the health care professions—are not incidental, but now structural in nature. The result of the ideology of technology will be that the growth of material prosperity will continue to pave the way in society for an increasingly individualistic, self-destructive society.

As well, the tendency to one-sided, materialistic, large-scale political frameworks is to be understood in the context of such an ideology of technology. And international politics are often deeply influenced by a kind of technical imperialism.

Is this a too pessimistic scenario? Time will tell.

The technical imperative reaps havoc

Under the influence of the ideology of technology, the chief normative values of technical development are *technical perfection* and the *technical imperative that states "what can be made, ought to be made."* Our culture has therefore entered a dynamic process of change supposedly leading us to a technical paradise, but in reality confronting us with a disastrous development.

Take ordinary jobs for example. The aim of raising labor productivity on the basis of the development of new technical possibilities leads to the human tragedy that many cannot meet the prerequisites demanded physically or psychologically and therefore drop out or are labeled incompetent and/or sick. Technical culture seems to cause burn-outs like smoking causes cancer. A high-tech society demands qualitatively highly educated employees who, due to the continually new technical advancements and rapidly changing organizations, continually must be re-schooled and forced to adjust to the new circumstances. Those who have a lower educational level remain unemployed, while the highly educated are more and more under pressure and often find themselves unemployed as well. The banking world is, at the moment, a good example of this.

Let's also not forget that this kind of development in the West demonstrably increases poverty and hunger in the Third World. This is because our technical progress and economic growth in the West (and the rich Eastern countries) are ego-driven, ignoring the duty of sharing the wealth with the poor around the world. And if the "Second" World (the former East European Communist countries) and the Third World are, perhaps, eventually raised by a back-breaking process to the same prosperity level as the West and rich East—and in itself, who would want to deny them that right?—the deleterious effects of the technical culture will take on world-wide, catastrophic proportions. Indeed, it is by no means an exaggeration to say that the future of the earth is at stake here.

The depth of the problem and Europe

Our own attempt at an *architectonic social criticism* (Kuyper's term) must begin by seeking to orient ourselves to the technological society around us. Due to an uncritical connection between science and technology, we are confronted with logical, inexorable consequences. Modern technical developments are obsessive, huge in scale, universal, reductionist, and socially leveling, and thus one-dimensional and impersonal. Human beings wind up being cogs in the machinery, depersonalized and alienated from each other. The end result is apparently rational, business-like, and efficient, but in fact an ice cold, barren culture, without a heart. Scientific-technical domination brings a tunnel vision and a tendency to massive scale to science, the economy, and politics.

The current spectacle of Europe trying to unite politically makes that clear. Sometimes it is maintained that the "irresistible power of technical advancements" make such unity inevitable. However, the trend toward a technically and economically strong, united Europe, far from reducing social problems, will most likely cause them to increase.

I believe we in a united Europe need a new commitment to the priority of sacrificial service to our fellow citizens and to the promotion of public justice. No more chasing after some golden, utopian, technological future, but rather a radical readjustment of political priorities is what we need.

A revised cultural mandate

We Christians must look again at the biblical cultural mandate. In Genesis 1:28 we hear God saying to Adam and Eve: "Be fruitful and multiply and fill the earth and subdue it and have dominion over the fish of the sea and over the birds of the heavens and over every living thing that moves on the earth." Even when we speak of such a cultural mandate for the glory of God and reject the current patterns of an economy where the rich get richer and the poor get poorer, if we neglect a critical reflection about technology, we run the risk of being infected by the malignant atmosphere of our age.

It is certainly true that the book of Genesis teaches us the cultural mandate, with its technical implications. But the Bible is very clear in many places that it is precisely technical developments which can lead humanity away from God. That is what we learn in the Bible from the narratives regarding Cain, the generation of Lamech, the building of the tower of Babel, and Nebuchadnezzar. In the last book of the Bible, the book of Revelation, the prophecy about the rise of Babylon is a very clear example, as well.

We ought to see the cultural mandate in the light of a wide variety of Bible passages, not just one. For example, listen to King Solomon's prayer for wisdom (1 Kings 3) and a life lived out of that wisdom:

> "He spoke of trees, from the cedar that is in Lebanon to the hyssop that grows out of the wall. He spoke also of beasts, and of birds, and of reptiles, and of fish." (1 Kings 4:33-34)

Here we hear Solomon speak of animals, not to subdue them, but to learn about and, in particular, learn *from* them. We see this regularly in the book of Proverbs. This is an aspect of the cultural mandate which has been neglected in the past.

Psalm 148, as well, transports us into another musical idiom, as it were, that we use when we connect the cultural mandate only with "subduing" and "developing" and thus often technical thinking:

"Heaven and earth, praise the LORD!"

A cultural mandate without such praise for the LORD as its constant expression is less than biblical.

Further, we see how Abraham, the father of all believers, is in essence a pilgrim. Abraham "was looking forward to the city that has foundations, whose designer and builder is God" (Hebrews 11:10). That special way of being and thinking, our pilgrimage existence, is seldom connected with the cultural mandate in Christian circles here in The Netherlands.

Strikingly and prominently, we see the Lord Jesus as the second Adam in Matthew 4 rejecting the temptation to serve Satan in order to make all the kingdoms of earth subject to Himself. This revises our vision of the cultural mandate in a fundamental way. For His attitude contrasts starkly with the way the cultural mandate is commonly interpreted to mean that we ought to exploit what we discover on earth in order to rule. And isn't this Jesus "the founder and perfecter of our faith" (Hebrews 12:2), Who took up His cross and thereby fulfilled the whole intended purpose of culture? What does it mean for us as Christians who claim to carry our cross, to be followers of this Founder and Perfecter of our faith? Isn't this precisely the core of what the cultural mandate means? Isn't it true that every cultural activity is therefore, at its root, devotion to and following Christ in faith?

The cultural mandate in history

Regarding the cultural mandate, historical questions are important. For example, why has the cultural mandate in Genesis, as it is usually understood by Protestant Christians, never been used by Jewish people to legitimize technical development? In the Bible, the Jewish people were dependent upon the technical skills of the surrounding peoples, including in building the temple. Scholars have pointed out that Western technical advancements, bearing the imprint of natural science, are not of Jewish origin, but derive from Egyptian and Babylonian culture.

It seems clear that Western Christians, in their conceptualization of the cultural mandate, have been influenced by the ideas of the anti-Christian Enlightenment, which, in its search to control and dominate through science and technical applications, made the autonomy of man its chief feature.

New Testament contours of the cultural mandate

The New Testament does not minimalize the importance of this earth and our works on it—including those of technology (Revelation 21:24, 1 Timothy 4:4,5) but it also calls us to focus the eyes of our hearts on the righteousness of the Kingdom of God which has come in Christ and is coming in the future. We are thereby called to carry our cross by denying the "works of the flesh":

> "Seek the things that are above" (Colossians 3:1)
>
> "For where your treasure is, there will your heart be also" (Luke 12:34)
>
> "For what does it profit a man to gain the whole world (that is, in our time, the aim of the ideology of technology, E.S.) and forfeit his soul?" (Mark 8:36)
>
> "Do not lay up for treasures on earth..." (Matthew 6:19)
>
> "Do not be conformed to this world..." (Romans 12:2)
>
> "For here we have no lasting city, but we seek the city that is to come." (Hebrews 13:14), etc. (Schuurman, 1989)

Connecting the divine world with our human world through Christ—hence by grace!—ought to determine our cultural attitude. Motivated by love for God, we turn away from our secularized and technical culture and seek to lovingly engage and serve our neighbors, nearby and far-off, and our own cultural setting.

The purpose of culture: the Kingdom of God

This provides a basis for a responsible stewardship in all the dimensions of life. Opposing the worship of the gods of culture, Nature, and materialism, Christians seek to be obedient to God. Technical development and economy are, in this way, liberated from a trust in progress and growth. Christians seek to follow a divergent, difficult, small-scale path of love, justice, service, self-denial, sacrifice, mercy, and thankfulness. These are keywords which simply do not belong to the vocabulary of the current technical worldview with its consuming interest in efficiency, effectiveness, and control. They call us to turn away from the ideology of technology, from the technical ideal of perfection, the technical imperative, and technical imperialism.

This conversion does not mean that we distance ourselves from all contact with or use of technology—we live in the midst of and through it and, honestly, we cannot live without it—but we may never give our heart to it, or live *for* it.

Is it possible to help shift the direction of our age? With God's help, we ought to try. If people begin to live out of the eternal perspective of the Kingdom of God, He promises His blessing. If the latter is a reality, then we can give technical advancement its justified but modest place. We begin to acknowledge the cohesion of reality, and the deep meaning and essence of things.

Our political responsibility

For some examples from real life, let's look at how politics in The Netherlands has been dealing with environmental issues. (Note from the editor; as a member of the Dutch Senate, the First Chamber of the federal parliament, from 1983 to 2011, prof. Schuurman has had first-hand experience in this area.) In general, Dutch politicians have come to believe that new technical know-how can help us. However, while technical means are partially effective in solving some of the problems of the environment, the intensity, the dynamic, and the increase of scale needed to apply new technology often undercuts the gains made.

This is demonstrable in the case of increasing cleaner greenhouse farming, a major industry here in The Netherlands. The individual farmer will imagine that his use of cleaner production is a contribution to solving the problem of environmental pollution. But when we analyze the current move from small scale operations to large scale ones in this sector, we see how the negative effects on the environment are continuing via the sources of energy used to heat the larger greenhouses, such as coal and natural gas.

The same is to be said about electric cars. They are emission free, but that is not so for the fossil fuel energy plants which provide the electricity. In the case of nuclear plants, other real dangers to the environment are present. With the current development of turning manure into usable methane, we see another danger. Pollution of soil and water is reduced, but the evil of reducing of animals to mere things to be manipulated remains intact.

Up till now, many politicians have spoken of sustainability as a priority, but little has actually been put into practice in The Netherlands. Without a change of the current trend of our culture, and a conversion to a new way of looking at the world, sustainability will remain a mythical concept.

That is why we need to be "converted" and reorient ourselves in a fundamental way. The result of slowing down and stabilizing the dynamics in our culture will mean that many of the technologies which have now been shown to be superfluous or counterproductive can be discarded. Technical development ought not to be allowed to function autonomously; it must remain a human tool, an assisting instrument, in service to humanity, to creation, and, ultimately, to God.

The real cost of things

We Dutch are notoriously cost-conscious. But how does this play out in practice? Here in The Netherlands, for both private and freight transportation, we have chosen to expand the highway system, putting down a lot of new asphalt in process, resulting in a new assault on Nature. In my opinion, it would be far better to let freight be carried underground, for example, on new tracks or roads under the existing highways. The initial costs would be high, but we would see a reduction of individualistic consumerism and create a new, healthy dynamism for the economy. Or am I dreaming?

Ironically, our technical applications cost *much too little* initially compared to the long-range, damaging, costly impact on the environment and society. Paying more for safer and more thought-through, forms of technology is really cheaper on the long run.

A new appraisal of the issues surrounding a new economic order without attention for the ideology of technology is doomed to failure. We must reckon with the long-time effect of a product's production and consumption on energy, raw materials, pollution, and the ecological system. Against the tunnel vision of a materialistic economy, we ought to shift to an economy where trees matter, a "tree economy" (the term comes from the economics professor, Bob Goudzwaard.) That image expresses ecological priorities with room for growth while assuring that the fruits of such growth are more balanced, proportional, and truly sustainable.

Schumacher still valid

I continue to agree with the well-known economist E.F. Schumacher (1911-77) (1973, 1977). Driven by the motivation of love for God, neighbor, and God's creation, we must strive for an ethical approach to technological development. The basis for this *ethical approach* lies in the concept of *justice as the cohesion of the order of creation and redemption*. This order is God's order. A culture which orientates itself to it is no longer anthropocentric, but theocentric. Living and building within this Divine order brings to light a great variety of norms, which all call for implementation (Strijbos, 1988, Schuurman, 1990).

The ethical frameworks, so brought to light, will restrain, in a healthy way, the development of information technology, biotechnology, genetic modification, and nanotechnology. It is true, the sectors in which responsible technical development should be stimulated, such as industry, agriculture, education, public health, and so on, will certainly not immediately see solutions to all problems. However, such problems will not easily get out of control, as they, at the moment, often are due to the hubris of a scientific-technical desire for domination of creation.

If such a new approach becomes a reality, even to a degree, the development of culture will be more stable and sustainable. All our neighbors (including those far off) and the earth itself will certainly benefit. At the deepest level, we will then be more able to live for God's glory and experience fellowship with Him.

That great aim remains our constant, driving motivation. We are praying and laboring until the great day dawns, the day of the great Restoration of all things in and through Christ, at His return!

Literature

Corbey, R and Van der Grijp, P., (1990), *Natuur en Cultuur: Beschouwingen op het Raakvlak van Antropologie en Filosofie*, Baarn: Ten Have.

Ellul, Jacques, (1980), *The Technological System*, New York.

Ihde, Don, (1985),"The Historical-Ontological Priority of Technology Over Science," in: L. Hickman, *Philosophy, Technology and Human Affairs*, Texas.

Kuyper, Abraham, (1911), *Pro Rege*, deel 1, Kampen:Kok.

Schumacher, Ernst F., (1973), *Hou het klein*, Baarn: Ten Have.

Schumacher, Ernst F., (1977), *Gids voor Verdoolden*, Baarn: Ten Have.

Schuurman, Egbert, (1985), *Tussen Technische Overmacht en Menselijke Onmacht*, Kampen: Kok.

Schuurman, Egbert, (1989), *Het Technische Paradijs*, Kampen: Kok.

Schuurman, Egbert, (1990), *Filosofie van de Technische Wetenschappen*, Leiden: Martinus Nijhoff.

Strijbos, Sytse, (1988), *Het Technische Wereldbeeld*, Amsterdam: Buijten & Schipperheijn.

Tillich, Paul, (1985), *The Spiritual Situation in Our Technical Society*, New York.

CHAPTER 3

FAITH, SCIENCE, AND TECHNOLOGY— SHIFTING FRONTS OF CONFLICT[1]

Introduction

The subject of the relation between faith and science has, deservedly, received a lot of attention from Christian university students through the years. This subject reveals a great measure of continuity, through the ages, but new aspects appear continually in the history of the Christian West. This is so because scientific thinking in the West was characterized by great, sometimes revolutionary achievements, and especially because Western philosophy, active in the background of science, began to turn against religion in general and Christian faith in particular.

Historically, more attention has been given to the relation between faith and (natural) science than to the relation between faith and technology. Tacitly, people seem to assume that the relation between faith and technology can be subsumed under the rubric of the general problem of faith and science; but there are important differences.

Many Christian academics have found their workplace at the university. This is certainly true today. There has perhaps never been a time when there were so many professors and lecturers from a Christian background active at the universities in The Netherlands. And, with some exceptions, it seems at the moment that Christians in these positions are no longer leaving the church in a massive fashion, as has been the case in the past.

However, at the same time, there has never been a time in which our culture, influenced by technical and economic developments, has been so secularized. Large groups of people, including many students, have left the church or are in the process of being estranged from it.

Are we, as Christians, facing changing fronts of conflict? If so, are we sufficiently alert to the new situation? I believe that the issue of faith and science needs to be supplemented by addressing the challenges of our technological age.

---

1 Edited version of a lecture originally given to a gathering of the Association of Reformed Students in Eindhoven, The Netherlands, in 2002.

Natural science vs. Christian faith

In The Netherlands and in the West in general (as well as much of the East), the achievements of natural science are highly regarded. It is maintained that such science has a monopoly on truth, that it is the key molder of culture, and that it is potentially able to solve all our problems. The claim that science has a monopoly on truth has always led to a collision with Christian faith and other religious convictions. The dominant view of science at the moment in the West is that everything can be explained on a materialistic, mathematical basis.

According to this commonly accepted narrative, God, His revelation, and statements regarding Him are excluded by scientific and philosophical criteria. There is no evidence for His existence and the suffering of living beings is sufficient proof that He cannot exist. Many, including many scientists and academics, openly express their belief that they cannot see a place for God in science. Even at theological faculties of universities, the resurrection of Christ from the dead is disputed, because (supposedly) scientifically a dead body cannot come back to life.

The Big Bang and Evolution

Almost unanimously, there is support for the theory that the origin of everything in the universe can be explained by the Big Bang, interpreted as an impersonal, random event without any ultimate purpose or meaning. Typical Big Bang proponents believe that some 10 billion to 20 billion years ago, a massive blast allowed all the universe's known matter and energy—even space and time themselves—to spring from some unknown type of (impersonal) energy. It then is obvious that the issue of the relation between faith and natural science must be focused on the question about the origin, the historical emergence, and the development of our universe.

As far as life on earth in concerned, Darwin's theory of the origin of species reigns, whereby all of life emerges as the product of pure chance and merciless competition in time. An evolutionary spirit permeates the academic world.

Every Christian has to do with this dominant, all-inclusive, intolerant narrative of origin in our culture. In my opinion, the "clarifications" of the standard Big Bang theory and Darwinian evolution are not the last words on these subjects by any means. I believe that we have to do here with a metaphysical overestimation of natural science. To exclude God either a priori or as a conclusion, is going beyond natural scientific categories and methods.

Learning from Galileo

Conversely, it is advisable for Christians to approach the relation between faith and natural science in a relaxed, open-minded way.

Looking back in history, Galileo's position was a strong one, seeking empirical evidence for theories about our solar system. On the basis of observable phenomena, he was correct in concluding that the earth revolves around the sun. We can learn a lot about being Christians in science from the case of Galileo.

Evolutionism and scientific creationism

What are we talking about when we talk about the theory of evolution and evolutionism? On the one hand, we can observe micro-evolution within species. Theories about this are legitimate. On the other hand, we are justified in making objections to speculative evolutionary ideas and rejecting evolutionism as an expression of a faith in naturalism. Naturalistic evolutionism has no place for God the Almighty, Creator of heaven and earth. Evolutionism is therefore a life-and-worldview, an extrapolation into metaphysics of a scientific theory.

If we say that we believe in creation and God the Creator, we are quickly labeled hopelessly out-of-date *creationists*. However, we ought not to commit ourselves too quickly, in reaction to evolutionism, to what is called *scientific creationism*, holding to a "young earth." This is a theoretical approach, which proceeds on the basis of the trustworthiness of biblical data and seeks to develop a scientific model of creation. The background of Darwinian evolutionism is, in my opinion, an over-estimation of the thinking capacity of human beings. However, when that kind of an absolutized thinking is continued when biblical data is accepted, the danger is to *presumptuously* theorize about scientific creationism in a similar fashion.

In the process of doing natural science, fundamental questions are to be asked of hypotheses and presuppositions, and doubts are legitimately raised about all too confident assertions. Scientists are always called to be *critical* scientists. They haven't always been so critical in practice. The spirit of rationalism in the West—within which natural science is absolutized and honored—has thought itself to be supreme during the last centuries. People thought that science could be carried on completely objectively and value-free. Science was thought to be able to be developed wholly independently of religion, faith, or a philosophical point of view.

In contrast, Christians were very much involved in science and academic activity during this time, but were conscious of their faith presuppositions. However, they were too little aware that they ought to take a distance from an absolutizing of thinking itself. It seems evident that both evolutionism and scientific creationism are still proceeding from a kind of presumptuous thinking which does not recognize its own limitations. Scientific creationism seems to believe—in other words—that we can penetrate to the core of created reality and its historical emergence, and describe it adequately. In fact, this does injustice to the miraculous character of the creation and the greatness and omnipotence of the Creator. It would be better if Christians studying at the university level would be satisfied with questions remaining unanswered, rather than attempt to find answers which are forced and inadequate.

Room for mystery, room for miracles

Christian academics can prevent themselves from becoming presumptuous by developing a sensitivity regarding the relativity of presuppositions in science and respecting the mysterious character of God's created reality. Then, science and spiritual fruitfulness can go together. Then a consciousness can develop which is aware that secrets will remain for the scientist which he or she cannot fully fathom. This process is beautifully described in the book of professor A. Van de Beukel, *Things Have Their Secret* (*Dingen hebben hun geheim*) (1994).

Creation cannot exhaustively or finally be described by human thinking. Scientific, academic knowledge is *limited* knowledge. Many scientists and academic researchers are reductionist, shutting their eyes to the miraculous reality of the creation and rejecting biblical miracles out of hand. Actually, it is part of the very structure of natural science, as a particular academic discipline, that it cannot say anything pro or con about those miracles. Natural scientific knowledge is abstract and universal knowledge. Miracles are unique events and have to do with God's special activity, not constant physical laws. The *miracle* of creation is revealed in the Word of God and can only be *believed*. The beauties and wonders of creation surround us and are open though never exhaustively to be investigated. Thusm science and academic thought ought not to forget their own limits!

Let's now look at various models which have been proposed regarding the relation of faith to science.

Faith the "crown of the sciences"

The first proposed solution to the relation between faith and science in its Western Medieval and different modern forms is what we may call the "crown of the sciences" view. The contemporary views of philosophy and science are independent of Christian faith—belonging, as they are conceived as being, to the terrain of Nature—are accepted, but supplemented or "crowned" with faith, the terrain of Grace. In the concept of Thomas of Aquinas in the Middle Ages, faith is then identified with theology. Actually this solution is, in fact, a synthesis of two conceptions, which each proceed from a distinct faith. Faith in an autonomous or self-sufficient science is rounded off with Christian faith, which is then elaborated into theology. Operating out of this conception, theology is called the queen of the sciences.

By the power of the church hierarchy, the tension between the two ways of knowledge—natural science and theology—could be contained for a long time. The church, with the help of theology, made the final decisions about the matters in dispute. The solution surrounding the conflict with Galileo may be seen as the most outstanding example of this. The church operated with a view about a geocentric solar system which finally could not be maintained. This mistake has, unfortunately, damaged the church's reputation up until today.

In the tradition of Roman Catholicism, this model of "theology as the queen of the sciences" is actually still in force. This synthesis between Nature and Grace has thus been sustained for a long time. Apparently, the ecclesiastical authority of the church guarantees it. The fact that there is decreasing support for it in some Catholic circles has not yet made major changes in how the church operates.

A faith witness supplementing science

In the Protestant tradition, this kind of synthesis thinking can also be found. In Protestant circles, theology, as the queen of the sciences, had the last word regarding scientific issues. Due to the lack of ecclesiastical power, the synthesis took on very different forms. But the synthesis itself remained important. Many Protestant academics tended and still tend to accept the presumably neutral current views of science more or less without criticism, and supplement this with a personal witness of faith. For example, scientific evolutionism is seen as unavoidable by many. People work with this view in their scientific fields, but as believers, they say that they continue to maintain their faith in the creation of all things by God.

This "supplement" in the sense of a "witness," is certainly sincere, but, in my opinion, this position has fundamental weaknesses. There is no real attempt here to *integrate* faith and science. A faith witness as a "sauce" over an "unbelieving" view of science or theory has little credibility. Such a strategy does no justice to faith, because scientific evolutionism, as we have seen, entails a *faith* commitment to naturalism as its basis. Philosophical and scientific thinking are developed wholly, and from the outset, outside the perspective of Christian faith. Adding Christian faith or theology to this picture, as the second story of the building of science and academic knowledge, remains a mere stop-gap remedy. This "queen" has no clothes!

Science and faith two different worlds

Next to the "theology as the queen of the sciences" position and, as its variant, the "faith-as-witness-added-to-science" approach, there is the popular view supporting a complete division between science and faith. The picture is one not of a two-story building, but of two separate buildings next to each other. Thinkers conceive of science and faith as *two independent terrains* of life. Next to the world of science, there is the world of Christian faith.

Unfortunately, this way of thinking can characterize our Christian life as a whole. Such a division occurs all the time in economic and political life, and in everyday life. We reserve a special day or special activities as being "Christian," and the rest of the time and in the rest of our activities Christian faith has nothing to say. In this way, Christian faith is effectively *privatized*.

This division means for the Christian academic that he or she lives in two worlds with two sets of truth which are, at root, in conflict with each other. The truth of academic study exists *over against* the truth of Christian faith. Whoever has an overview of history in the West knows that this struggle is often decided in favor of science—or better—in favor of *faith* in (naturalistic) science. This is so because we are looking for certainty in that which is based on proof, that which is logical, that which is authoritative, that which is necessary, and that which is unavoidable according to the current dictates of science and academic judgments.

A clear example of what I mean concerning a division between faith and academic study is demonstrated by the oldest Dutch universities. They all began as Protestant universities. Now they have become humanistic, non-christian bulwarks. Western philosophy has been dominated by the driving impulse of the idea of human autonomy since the Middle Ages. And it is this philosophy which, with idolatrous power, has worked and is still working in the background of every academic discipline from mathematics to theology. Laplace as a mathematician even believed that we could eventually predict the future by mathematics. And the theologian Rudolf Bultmann denied, on the basis of natural science, that God would ever break through the existing order by miracles or even by His Son Jesus Christ. This secularizing of the enterprise of science and academic study has led to the secularizing of the entire culture via the convincing cultural power of science. The total separation of science from Christian faith has led nowhere. This "queen" has left the building!

Foundational thinking

A third variant of the relation between faith and science is found in "foundational thinking." It is maintained, correctly, that as Christians, we must affirm an integration of faith and science from the outset. That's why Christian principles should formulate which together form the basis for science and academic study at the university.

This view seems very attractive. Nevertheless, there are a number of problems with it, in my opinion. For example, what exactly are the principles we're talking about? Are they really foundational? The translation of God's Word revelation into scientific and academic principles is not so easy to do as it seems. Before you know it, you've chosen certain starting points and made everything dependent upon them. That happened, historically, in the case of the Free University of Amsterdam. An attempt was made to clearly formulate and stipulate the Scriptural or Reformed principles for carrying on science and academic activity, but these principles later turned out to be questionable and even untenable in practice. Due to this approach, the ideal of Christian scientific and academic endeavors at the university level was eventually discredited in The Netherlands.

A better way to do it

It would have been much better if more attention had been given to the special character of the Christian foundation of science and academic research. Of course, one can say that *thinking* in the formulation of the foundation is a mediating activity and that we are to build on the foundation by means of that thinking. But what is the relation between that thinking and faith? There is a danger that we regard faith as something static, which is to be formulated in certain statements and rules, and then as result assume that the thinking which follows these statements will be sound. In other words, we concentrate on the presuppositions, and believe that the differences are to be found between Christian and non-Christian science and academic activity. I would say there, at the level of presuppositions, we *do* see basic differences, but the challenge remains: For what are the consequences of faith for thought and research, both in the formulation of principles as well as in the thinking which is to follow, based on that foundation?

Earlier in this lecture, I tried to make clear that in the case of scientific evolutionism and scientific creationism, however different their presuppositions, both are imbued with presumptuous thinking regarding what scientific theories can achieve. Both have unfruitful, "fundamentalist" tendencies.

Better: faith regulating thinking

It has become increasingly clear to me that the issue of faith and science is usually wrongly formulated. It is not the issue of the relation between faith and science as such, but the contrast between *Christian* faith, within which there is room via a Christian life-and-worldview, for a certain view of science, and a *faith trusting in the currently accepted scientific theories*. Faith and thinking are always involved with one another. The question is, *which faith directs or leads the thinking?*

In Western culture, the struggle between these *two faiths* has been going on for a long time. Christian scientists, academics, and Christian university students are continually involved with it. They are forced to try to define their own position in the struggle. More often than not—and mostly unconsciously and in good faith—they adopt one of the synthesis options. Such a synthesis is, however, never the end of the story. Within the synthesis, the struggle between two faith commitments continues, whereby the faith-in-atheistic-science commitment all too often wins at the cost of Christian faith. If Christian faith is to define and structure scientific and academic activity, then scientific and academic activity must be carried *out of* Christian faith and *for the sake of* that specific faith.

The promise of Reformational philosophy

In the Reformational philosophical tradition, in the footsteps of H. Dooyeweerd (1894-1977) and D.H.Th.Vollenhoven (1892-1978), much attention has been given to examining the relation between Christian faith and the various sciences and academic fields studied at the university. In that tradition, we presuppose and philosophize in the divine light of revelation, which reasoning ought to proceed, and out of which a vision of science and academic activity arises.

Christian faith and scientific-academic thinking are two differently qualified activities of human beings which cannot be derived from each other. What Hebrews 11:1 says about faith cannot be said of scientific-academic thinking: "Now faith is the assurance of things hoped for, the conviction of things not seen." But this fundamental distinction between faith and thinking does not mean that they exclude each other. Of course, "I believe" and "I think" are activities of one and the same person. In the "I-ness," in the *heart* as the religious center of every human being, the coherence between believing and thinking (and acting, thus making technical applications) is concentrated.

The heart is, biblically speaking, the center of the human person, upon which God has imprinted His seal—since to be human is to bear the image of God. Out of the heart flow the springs of life, the Bible says (Proverbs 4:23). That which possesses the heart in a religious sense, which constitutes its anchor in life, gives shape to all its expressions, such as believing, willing, thinking, etc. The center of human beings is thus not *thinking* by itself as is believed in the Western philosophical tradition, but the *heart*. And, flowing from that heart, there ought to be a unity between one's faith and one's head, unity between faith and scientific-academic activity (and thus also technology).

It is quite evident that, in practice, this unity of heart and head (and hands) is difficult to achieve and maintain. More often than not, our heart is divided. It is so easy for us as Christians to fall back into a split life, that is, *different faith commitments compete with each other for supremacy within us*. This phenomenon calls for some elucidation.

Believing is active in the background of everything we do. When we witness to our faith, it becomes evident that the religious content of our heart is where our heart finds certainty and stability. The content of faith directs our concrete activity, practical life, and non-scientific, non-academic thinking and knowing. Obviously, pre-scientific, pre-academic thinking and knowing is inherent to faith, but faith transcends the latter. Scientific-academic thinking, in turn, presupposes non-scientific-academic thinking and knowing, and is furthermore also carried along and stimulated by trust or faith. That basic trust is the center of human experience. In other words, *believing* directs or regulates experience, practical living and thinking, as well as science and academic activity. Led by faith, of different kinds, all human activities are opened up and made potentially fruitful. No one lives without faith.

But there is also a flip-side. Let's limit ourselves to science and academic activity. In Christian faith, we discover the mandate and the final boundaries of science and academic activity. Faith-in-science does not acknowledge such boundaries. That is the reason why we often see that the limits of science and academic research are exceeded and we find ourselves in the realm of groundless, sometimes dangerous speculation. That is so, as we saw, in the case of scientific evolutionism.

But in the case of Christian faith, fed by God's revelation and enlightened by the Spirit, faith leads us to perceive and accept what we can discover in God's creation, as well as normative boundaries for our own thinking. Science and academic activity are carried on in a faith perspective with our purpose in life in full view. Science and academic activity are a mandate given to humanity and a possibility given by God to human beings to investigate and understand, in a creaturely way, all things. This means that new, sometimes very surprising things in creation and history can be discovered. This is not in conflict with Christian faith. To the contrary, biblically grounded, normed science and academic activity enrich faith! The scientist and academic can enjoy the possibilities of doing science and academic research to the glory of God the almighty and all-wise Creator.

The alternative: faith in human autonomy

Currently in our Western academic tradition, we see a faith in human autonomy gaining strength. Since the time of the Greeks, such faith has been connected with philosophical or scientific-academic thinking. Concretely knowing and understanding reality centralized on a basic trust with the heart, is increasingly replaced by scientific-academic thinking in abstractions and logical systems. This leads to what we might call a "scientification" of ordinary human life, and finally to a worldview that is closed off to God Himself.

This basic characteristic of current Western thinking has strongly influenced the search for truth. Truth became a *theoretical, scientific-academic* truth. This theoretical truth, inspired trust due to its claim to universality. Human beings projected their idea of autonomy and their own greatness in terms of science and academic discourse. The relativity of all scientific and academic pronouncements was lost. Absolute claims triumphed. People gave themselves over to the compelling, logical, and apparently incontestable character of scientific-academic knowledge.

We call this chief trend Western *rationalism*. It is characterized by theoretical, logical, and abstract truth. Certainly, abstractions belong to science and academic discourse. Such abstractions become risky only when people don't realize that they are constructing abstractions. Theoretical "truth" as an *absolutized* theoretical truth is at odds with a faith that acknowledges that Jesus Christ is the Way, the Truth, and the Life.

Faith in Christ from first to last

As Christians, we believe that all things have received their existence through Christ. All things, now deeply stained by sin, are reconciled again with God and receive their ultimate purpose and meaning in Christ. This truth is much richer than a scientific-academic theory. It demands full personal involvement, a self-surrender in trust, willingness to listen and then say: *I believe in Jesus Christ!*

The content of our faith from God as Creator to God as Consummator is not a matter of logical proof. "*By faith* we understand that the universe was created by the word of God...." (Heb. 11:3). This is not something we can *comprehend*—that is, logically or scientifically adequately conceptualize—but certainly is something we should *believe*.

There is no *fully logical harmony* possible between thinking and believing. The theme of faith and scientific-academic activity and the relation between believing and thinking cannot be worked out by thinking as such. If that were so, then scientific-academic thinking would rule over faith and would dissolve the mystery of faith.

Faith demands that, at the boundaries of scientific-academic thinking, God and His activity should be acknowledged as a mystery. Faith accompanies science and academic activity from the "other side of science and academic discourse," out of faith and for the sake of faith. Faith regulates or opens up scientific-academic thinking and establishes its limits by which science and academic activity retain their own meaningful place.

We are to take every thought captive for Christ (2 Cor. 10:5) the

King, including in the realm of science and academic activity. This is part of the true renewal of our thinking (Rom. 12:2). Christian scientists and academics know that they are called to be active in believing, to be wise, careful, patient, and thoughtful. Everything which does not proceed from faith is sin, the Bible says (Rom. 14:23). That is, without faith ,the things we think and do, including science and academic activity, do not fulfill their special calling and purpose. Thinking must be continually renewed under the leading of the Holy Spirit and by a living faith (Ef. 4:23). Scientific-academic thinking must be rooted in and borne by a practical wisdom for living, of which Jesus Christ is the source and the root. Through faith, an inward, deep relationship with Christ ought to motivate Christian thinking. The purpose of science and academic activity is to glorify God.

Discoveries are made by all

This purpose and meaning of science and academic activity can be perceived even when a particular scientist or academic doesn't acknowledge God. When I read the weekly supplements on science and academic research of the NRC/Handelsblad (a Dutch non-Christian daily newspaper, ed.), I am often filled with surprise and admiration. As a reader I observe, in general, a human haughtiness which new scientific developments seem to stimulate. But even a misplaced faith in science always remains bound to the contours of God's creation, and therefore unbelieving scientists very often bring to light "states of affairs" which we can certainly appreciate and value.

It is clear that the discoveries and formulations of science and academic research can be ultimately integrated with Christian faith. When we look at the results of science and academic activity with the eyes of faith, our amazement and gratefulness grow.

The shifting front of conflict: relativizing scientific-academic truth

It is not the case that everyone in our day is a rationalist. People have come to see that there are too many uncertainties involved with scientific and academic knowledge to remain unreconstructed rationalists. For example, the internal developments in the area of physics, with Heisenberg's uncertainty principle and Einstein's theory of relativity have shown us the untenability of making final truth claims in science.

Seeing the influence of certain external factors in the construction of scientific theories has also contributed to tarnishing the image of final scientific-academic truth. Thomas Kuhn has demonstrated with his analysis of paradigm shifts in science that historical and social contexts have shaped scientific theories. The philosopher Feyerabend has gone so far in pointing out social factors shaping science that people call him the anarchist philosopher of science.

The rise of "system thinking" has created space for a pluralism of method in science and academic research. Not just the analyzing and reductive method, but also the extrapolating, "expanding" method is now accepted in general. Hereby an enormous flexibility has been introduced into the area of forming scientific theory. The rigid thinking of a cut-and-dry rationalism is being challenged. At this point, we see how one front of conflict for Christian scientists and academics has shifted.

Some go so far as to question the very trustworthiness of scientific-academic knowledge itself. Christians ought not to go along with this irrational turn, while at the same time taking advantage of this new uncertainty.

A Christian vision of science and academic activity is now a real option in the marketplace of ideas. I think that we may now expect that Dutch universities will be more open to the contribution of orthodox Christians. Ironically, this will probably be less so for philosophers and theologians, the majority of whom remain rationalistic, but the climate definitely seems to be more friendly to Christian natural scientists and other academics. Recently, various Dutch universities have had presidents from an orthodox Protestant, Reformed background, and I can count more than 100 professors and lecturers who have this background.

From my point of view, it would be great if a consciously Reformational view of science and academic activity could gain a hearing in this situation. To be honest, at the moment, this approach is more tolerated than appreciated by many Christian academics. There is work to be done!

The shifting front of conflict: from faith in reason to faith in control

While the idea of the final truth of scientific and academic theories may have become suspect, other presumptuous claims are popping up. Many are convinced that, while final truth is impossible, we ought to strive to *manage* and *control* reality. The aim of determining the *truth* of science and academic research is shifting to its *usefulness*. We can call this a *pragmatic turn*. Here we see the front of conflict shifting increasingly to the area of technology and economics.

Christian university students ought be aware of the veneration of science and academic research verging on idolatry. But at this moment in time they are more likely to collide with a whole society, including the university, which has become totally enchanted with technology. We are under its mesmerizing spell! The autonomy of enlightened scientific-academic thinking has become physically tangible in the technical achievements around us. We can call this a *scientific-technical*, or *technical culture*, which is also a strongly *secularized culture*.

This godless culture forms a second front of conflict for many young students. Perhaps the confrontation with faith in reason is simpler because, in a certain sense, it involves a criticism of rationalism's irresponsible claims. Criticism of faith in technological management and control is, on the other hand, especially self-criticism because it defines in every way our own all too prevalent *cultural attitude*.

The shifting front: from a faith in control to technical thinking

At the moment, in the West and increasingly in the East, human beings see themselves as "Lords and Masters" (the term is from the father of modern philosophy, Descartes) who make use of technical rationality for certain ends. Particularly the natural sciences and the engineering sciences are exploited with the ambition of bringing everything under the sway of humanity. The desired result is comprehensive solutions to old and new problems of society. The ideas of Francis Bacon have been particularly significant, as they kept up the hope that by scientific, technical progress a return to a lost paradise would be possible.

The ambitions have increased faith in technical control. In a certain sense, only those problems which are susceptible to being solved through technological applications are acknowledged as real. Deep reflection about life and religious problems is tacitly avoided. We see around us and in us a spiritual emptiness on a scale which is unprecedented compared to the past. The fact that this is not widely acknowledged makes the situation even more serious. The result is that this technical style of thinking or mentality suffuses the entirety of our culture.

We've got a problem here! Let me give two examples.

In the field of biotechnology, (from the Greek word for life, *bios*) it is ironic that justice is not being done to the phenomenon of life itself. The technical model of life fails to appreciate what life in all its complexity and depth really is. That's why it is not surprising that biotechnology (including genetic manipulation or modification) has to negotiate so many hidden pitfalls.

Another problem of this moment is that some think that the biggest danger of the Internet is the so-called spreading of disinformation and lies. They think that the ethical problems will be automatically resolved when Internet is "cleaned up." However, this overlooks the fact that the current technical mentality stimulates a certain kind of reductionist behavior, reducing life to computer code, to bytes of information. Cleaning that up is considerably more difficult!

When human beings are preoccupied with technical applications 24-7 their entire being is eventually possessed by computerized data. We lose our sensibility to the many other dimensions of life. Paradoxically, a gigantic increase of information is the cause of a decrease of meaning and significance in our society. The more interactions there are online or on social media, the less real contact we have with one another, the less we are spiritually in tune. Therefore—we say it again—we see the rise of a technical mentality go together with a growing secularization.

A vision concerning technical development

We should expect that Christians involved in developing new technologies would be aware of God's call to care for the earth. They, of all people, should have been alert to dangers of the apocalyptic disasters such as at Chernobyl and Fukushima, or the oil disaster in the gulf of Mexico. They, of all people, ought to be sensitive to sustainability issues and should be attracted to a non-luxurious, or even ascetic lifestyle. Calvin's call for simplicity should help us to live for our Creator.

Let's not underestimate how deeply the developments in natural science and technology affect the way we all live and think. Being human, seen from the perspective of creation and also in the light of sin and grace, is a constant in history. But being human, in our experience of being human, is also defined by our historical context, including the tools and instruments we use, and the culture which we breathe in every day. When our Western culture, in this electronic age, is preoccupied with building a modern tower of Babel, then we as Christians are either inhabitants or at least close neighbors of that tower. We can't avoid it.

The big question we face for the future is, how will Christians position themselves within this Babel culture so we can contribute to genuine cultural transformation?

We are called to help people see the limits of science and academic thinking and to give needed attention to a normative way of framing and approaching technological thinking. In this way, we can do justice to science, academic research, and technical developments and we offer in the following chapters of this book what I believe to be a perspective on a *salutary and fruitful* future.

Literature

Van den Beukel, Arie (1990). *De dingen hebben hun geheim. Gedachten over natuurkunde, mens en God.* Baarn: Ten Have.

Schuurman, Egbert (1985). *Tussen technische overmacht en menselijke onmacht–Verantwoordelijkheid in een technische maatschappij.* Kampen: Kok.

Schuurman, Egbert (2003). *Faith and Hope in Technology.* Toronto: Clements Publishing.

CHAPTER 4

LIBERATION FROM THE TECHNICAL WORLDVIEW: A NEW KIND OF ETHICS[1]

My personal story

During the 32 years that I have given courses on behalf of the Reformational Philosophy association, I have devoted my energies to treating modern philosophical movements, philosophy of science, and philosophy of culture with an emphasis on the significance of technology. Since being appointed to this position, it was natural that I would also give courses on the *ethics of technology* from the standpoint of Christian faith.

The ethics of technology has been increasingly addressed by others active at universities in The Netherlands. Influenced by the *Position paper on ethics and scientific research* of the then Dutch national minister of education, Ritzen, in 1986, the subject of the ethics of technology within the Technical Universities has received mounting attention. At Delft University of Technology, reports have appeared,[2] the Dutch Royal Academy of Sciences has devoted attention to the subject,[3] and, starting a few years ago, the University of Eindhoven has done so as well.[4]

The result is that, starting a few years ago, courses on ethics have been required of all students in Delft. In Eindhoven, this path is also being followed on a trial basis. Furthermore, platforms for the study of ethics and technology have been established at those universities. This is all very understandable. Engineers, in their capacity as researchers, designers, developers, and managers of technical systems, are having more and more to do with ethical problems.

When courses on ethics were made compulsory in Delft, they

---

1   Edited Concluding Lecture, May 15, 2002, at Delft University of Technology.

2   The Ethiek en Techniek committee, *Deeladvies I: Onderwijs*, Delft University of Technology, June, 1994.

3   De Forumrol van de KNAW inzake ethische aspecten van wetenschappelijk onderzoek, Amsterdam, 1994.

4   Onderwijs in Techniek en Ethiek, report of the Governing Committee Technical Applications and Ethics, February, 1999.

asked me (in a friendly way) if I might be able to forgo my course on ethics. In Eindhoven, I received the same request. The result was that I decided to emphasize with even greater clarity that I offered courses on the ethics of technology from a *specifically Reformational Philosophical* perspective and an agreement was reached. I was glad for this because I believe that I discuss things which do not get sufficient attention in the regular university courses. There, the emphasis is more on descriptive ethics.

For example, in the first Delft report, and in the course text itself, we read that in treating ethics no normative standpoints ought to be presented as authoritative.[5] Here there is only an analysis and evaluation of practical examples. World-and-life-views are, at the most, presented in a cursory, descriptive way, within this approach. A vague, bland, broadly acceptable approach to ethics is presented. But it is clear when conducting a deeper analysis that the ethics presented reveals underlying points of conflict which can be traced back to convictions stemming from different world-and-life-views. A Reformational perspective is needed here!

It is reassuring to see that, in this situation, while there are different approaches presented by different lecturers at these universities, there is agreement as to the challenges of the common problems. Furthermore, there is a general awareness that there is a lot of common ground in dealing with the issue of technology and that we ought to seek to achieve truly beneficial, long-lasting results for our common society via technology.

Beneath the surface

Current technical developments have fulfilled many initially promising conceptions and there are many potentially promising developments for the future. We can, however, be hypnotized into a state in which we have no real sense of the actual dangers which may follow. How is that possible? I have a hunch that the reason for this is that in our current Western culture, we have a built-in attitude of finding everything fantastic that technology may offer. In line with the spirit of modernism—and postmodernism—people find it difficult to reflect critically about such magical inventions. It seems that at the core of modernity, there is a kind of silent worship of technical advancements.

Granted, technical applications are often fascinating and exciting. But there is another side to the story.

The current monopoly of a technical worldview regarding reality means that everything is seen through a mentality of control and man-

---

5   See Ethiek en Techniek, course text by H. Zandvoort e.a., p. 7

agement. Through technology, we appropriate reality more and more but at the same time, we are estranged from it.

That problem manifests itself concretely, not just in the pollution of the environment, but also in tending to reducing existence to numbers and images on screens resulting in deep social tensions and conflicts. Despite ecological, green counter-reactions, there is still a widespread uncritical appreciation of technology. That's why ethics are so needed in this area. We read about ethics for the environment, ethics of economic development, or business ethics, but seldom about ethics of technology, which, I believe, is foundational for all the rest.

In this context, we in the tradition of the Reformational Philosophical need to address the challenge of formulating an ethics of technology.

An ethics of technology is needed

Modern technical developments, formed by science, have historically emerged from the Judeo-Christian tradition in the West. Does this tradition have something to say about a responsible ethics of technical applications? This question is urgent, since the Judeo-Christian tradition has been criticized as one of the chief causes of the destructive results of technology.

For the past 1,000 years in the West, classic technical craftsmanship, in itself a specific, though limited, cultural form, was characterized by its inter-human relationships and human scale. In general, the negative effects of such craftsmanship were manageable. Furthermore, this technical craftsmanship did not mold and structure the culture through which it was exercised, but was merely one part of it. It was embedded in the natural order, as it were. In this relatively static situation, there was no need for a specific ethics of technology.

1. The new "technical culture"

Compared to a century ago, we find ourselves in a totally new situation. Modern technology has developed in a stupendous way. It has permeated and now structures our societies as a dynamic technical movement, and has become a world-wide system. With modern technology, everything has to do with everything else, everywhere; it is our common technical milieu. Without modern technology, our human existence on earth would collapse.

The connection between technology and world economic life has undergone a tumultuous and powerful development. They are now strongly intertwined with one another. Thus, while I'm asking for attention for an ethics of technology, the latter cannot be considered completely divorced from an ethics of the economy. Both are joined together by a technical mentality—a technical framework or mode of thought.

Underneath the euphoria surrounding technology, a lurking uncertainty can be discerned. This is so because we have no experience with something like it before in human history. That makes the attempt to formulate an ethics for technology extremely challenging.

2. Advantages

When we compare our age with that of a few centuries ago, we see that modern technology has brought us, in the West and to a growing extent in the East, great advantages. Through it, the average human life expectancy has increased. Sewage systems and water purifying systems provide us with a healthier environment. The processes of mechanization, automation, and the use of robots have rescued people from heavy physical and routine labor. The connection between the economy and new technology has led to a level of unequalled material welfare. The Internet supplies us with an endless supply of useful information. Our wildest dreams seem to be fulfilled before our eyes!

3. Technological nightmares

However, in our contemporary world, the dark side of technical development is becoming clearer by the minute. Technological nightmares keep us awake at night, or ought to. Let's list some of them:

- All over the world, we see the threats of nuclear weapons (North Korea!);
- Radioactive waste has been and is continuing to be stored in all kinds of vulnerable locations;
- Arable, fertile land is becoming desert on all continents;
- Everywhere, we see a loss of food production;
- Raw materials are being overused and consequently disappearing;
- Many plants and species of animals are extinct or going extinct;
- The forests of the world are quickly disappearing;
- The oceans are gradually filling up with plastic;
- The fresh water bodies are salting up;
- The ozone layer continues to be depleted;
- Air pollution is increasing everywhere;
- The warming up of the earth's atmosphere seems definitely to be mainly man-made;
- Genetic modification of living things—plants, animals, and even human beings—is increasing at an alarming rate without any guarantees of safe consequences for life itself on earth.

- In the midst of these problems, the Internet and mobile phone use, while having many beneficial uses, are the cause of a reduction of "face to face" contact between people, whereby mutual estrangement, loneliness, and social disintegration are felt.

All this suggests a scenario more dystopian than utopian!

4. Vulnerable technology

And there is more to be said. Large-scale technical developments have proven time and time again to be vulnerable and risky. Due to human error or badly functioning technical applications, we are confronted with far-reaching consequences. Examples are, among others, the case of Chernobyl, the chemical disaster in Bhopal, various destructive computer viruses, creating havoc, and so on. And, naturally, the terrorist attacks which continue to plague our societies have made more than clear how vulnerable our Western culture is, because, among other reasons, it is so dependent on modern technology.

5. Technology itself perhaps the greatest threat

Our conclusion is that while in the past humanity was threatened above all by the forces of Nature, now we are above all threatened by modern technology going out of control. According to Einstein, "Technological progress is like an axe in the hands of a pathological criminal." How, if at all, can we regain control over all these technical applications gone astray? What is the right way to proceed? Ethical challenges enough!

What is ethics?

What do we actually mean by ethics? We may say that the subject of ethics as an academic discipline entails reflection about what constitutes good or responsible activity of human beings. In order to construct an ethics of technology, we must discuss what is good or responsible human activity in and through technology.

I see ethics not as one separate academic discipline, but as calling for a multidisciplinary or interdisciplinary approach. It concerns the activity of human beings in giving a response to all the normative aspects of reality.

Opinions diverge when answering the questions about what being human means, what is good and responsible, and what the place of technology is in our world. At issue are the grounds for ethics, differences in philosophical vision, and differences in life-and-worldview. These differences make discussions about ethics more difficult because at the bottom of things, there is no unanimous vision or even a working consensus to be found. We have to do with a plurality which scatters rather than unites.

Or is there, in spite of all of this, a common, central path in the cultural, historical background of these various visions to be discovered?

Background

Generally, in discussing the problems and threats of the Technological Era, people limit the discussion to surface phenomena, to the symptoms. We need a deepening of reflection in order to reveal the underlying historical developments.

I call the present dominant thought patterns of Western culture a *technical mentality*. When we pay attention to its historical background, we see that this mentality has deep roots. In the course of time, influenced by this technical mentality, belief that reality is created—and thereby faith in a Creator—has receded from view. Historically, the technical mentality flows from belief in the autonomy or self-sufficiency of the human thinker.

Let's look (as we have done in a previous chapter of this book) at Francis Bacon (1561-1626). He has been called the "trumpeter of the modern age." With expressions such as "knowledge is power," and "Nature, to be commanded, must be obeyed," he anticipates a situation in which, technically speaking, everything we want will be completely realizable. Nature is forced to serve humanity and therefore become our slave. Bacon says that humanity can eventually explain all the processes in Nature if we can only acquire full insight into the hidden structure and secret workings of matter.

However much Bacon tries to put a religious, Christian cast to this enterprise, we must agree with Hooykaas when he claims that Bacon is driven here by pride and hubris. In his utopia, the *Nova Atlantis* (the new world) Bacon describes an ideal society where all the power is in the hands of scientists and engineers. *They* make true Progress possible! He says that the human development of science and technical applications must be interpreted as an imitation of the divine works of creation. Biblical, eschatological perspectives are turned into expectations of future earthly, human progress. Bacon was even of the opinion that we can overcome the results the Fall into sin by means of science and technical applications. This redemptive motif is just as characteristic of technical thinking as the creation motif.

We move on to another 17th century philosopher, Descartes. He can be called the father of all of modern Western thought. Descartes sees rational thought as making use of the natural sciences and the technical sciences as instruments.

The ambition and expectation is to make reality subservient to human beings and therefore amenable to finding solutions to all the problems we face.

Descartes says that the laws of mechanics are the same as the laws of Nature. He sees Nature as a congregate of machines and a series of mechanisms. That's why the mechanizing of the West's worldview makes a breakthrough with Descartes, according to Dijksterhuis. Descartes says, "Nature is a machine, just as easy to understand as clocks and automatic mechanisms, when we look closely enough at it." Consequently, when we know the way the powers of Nature function, Nature can be calculated and guided. Because, for Descartes, humanity is the "maître et possesseur de la Nature" ("Lord and possessor of Nature") and we can learn to rule over Nature and make it obey us.

Descartes does not see plants and animals as having their own indelible, unique character and integrity, but, in fact, as things, machines to be manipulated. Descartes' idea is that via manipulation we will, one way or another, be able to make beneficial use of these "things." Reality as a reality to be manipulated is seen as being, above all, useful for humanity. Reality thus turns eventually into the technical use which human beings make of it. This conception of Descartes we see later on—in our own time—as it is applied in the bio-industry and in the development of genetic modification and manipulation. The technical mentality is always hungering for more territory to conquer. The 20th century philosopher Spengler expressed this pithily when he said, "The use of technology is as eternal as God the Father. It redeems life as the Son, and sanctifies it as the Spirit."

The genie out of the bottle

Technical thinking, once it becomes dominant, is inexorable. Once the genie is out of the bottle, it is unstoppable. It does not recognize the impenetrable secrets which at the deepest level characterize reality as God's creation. Technical thinking, always restless, continues to reconstruct the totality of reality. Indeed, the idea is that within reality, seen as one big machine or great information system, everything can be measured, weighed, counted, and technically managed and controlled.

It is therefore not surprising that recent technical culture also goes together with secularism and spiritual emptiness on an unprecedented scale. We could say that behind the mask of modern technical applications and autonomous, individual freedom, a spiritual emptiness is lurking—a "nothingness" which is the modern "ghost in the machine." The fact that this is not acknowledged makes the situation even more serious.

Inspired by the successful development of natural science, the heroic, Promethean humanity of the Enlightenment imagines that we can overcome all obstacles and renew ourselves and society. Because no other norm beside that of instrumentalist science is acknowledged, the way has been opened to attempt a limitless scientific-technical manipulation of the world. This totalitarian role of scientific thinking means that every non-scientific authority is declared incompetent. Here we see a definitive break with God as the Source of all things.

In a certain sense, this whole historical process finds its climax in that which—using Habermas's terminology—can be called the ideology of technology. This ideology wears self-chosen blinders which block a full view of things. Questions regarding fundamental, essential issues are eliminated as being irrelevant, such as questions regarding the backgrounds of technical developments, questions about the origin, the purpose, the driving motives, and the values and norms for technology.

The religion of technology

We can find some striking examples of an expected future technological "heaven on earth" in David Noble's book, *The Religion of Technology*. He shows that, since the Renaissance, thinkers have held the audacious ambition that humanity can become a kind of god in and through technical developments. For the first time in history, technology is being connected to the idea of being a co-creator and a co-redeemer with God Himself. In spite of the continuous impact of evil in the world, people in philosophical and scientific circles believe that an original paradise can be restored via technology. The technological humanity is the new Adam. Starting with this view of things, the religion of technical development becomes a replacement for the Christian expectation of the future, the Kingdom of God. According to Noble, this technological "heaven on earth" expectation is active in all the newest areas of technical development. He is not making up stories here. He shows the religious ardor of people by means of quotes from space experts, those doing research in artificial intelligence, developers of cyberspace and virtual reality, and representatives of those involved in genetic modification. The boundaries of space and time are crossed; human beings seek immortality as machines and a perfection of the mind, not bound by the body, which will be omnipresent in the cyber age. Genetic modification guarantees a recreated, new humanity.

The technical worldview

Due to the absolutizing which characterizes the technical mentality, the core and substance of reality is lost. Whatever doesn't fit in the technical model is ignored or forgotten. Our Western culture has now become addicted to this mentality. Every new technical development—all the new discoveries and innovations—makes this worldview more dynamic and increases its hegemony.

But, in fact, it remains a construction of human beings and functions as a cultural paradigm. It is indeed a kind of ethical framework within which people think and act. It has a normative significance; motives, values, and norms are derived from it. That which is determined by natural science and can be technically made, is, as it were, the true reality. It has increasingly suffused and colored the development of Western culture, and stamped and structured the current globalization as well.

This worldview has thus been derived from technical developments and has an influence far outside the realm of technical applications. Not only our relations to the environment, but our whole human society is shaped by it. All of us breathe in the heady, addictive air of this *technical mentality*.

The current ethical approach inadequate

The technical worldview is not only the hidden cause of the problems we have highlighted earlier, but, ironically, frequently motivates the search for ethical solutions to our problems. The usual, current approach limits itself, for the most part, to advocating a behavior of avoiding risks by acting in a controlled and cautious fashion. I have sometimes called it a supremely technical form of ethics. Ethics become a series of naive techniques. This "ethics of management and control" focuses on avoiding or removing undesirable possible symptoms, while the huge technological, apocalyptic mess we've put ourselves in (remember our list above) is hardly noticed.

For the *technical mentality*, the first and greatest commandment is, "be as effective and as technologically advanced as possible," and the second is like it, "be as efficient and cheap as possible." On these two commandments depend all of technical, materialistic culture (paralleling Matthew 22:35-39).

We are caught in the web of technological thinking. As we continually seek to adapt to the existing technological predicament, it becomes more and more difficult to look for a different fundamental vision of reality with truly different motives, values, norms, and ethical moorings.

Sustainability not in the picture

What about the environment and earth, our home?

Carried along by technical thinking, material values and norms also dominate our culture through the theme of *progress*. Problems of the environment and of Nature are acknowledged, true, but mainly addressed tangential to the theme of *human survival*.

Here the technical worldview and the ethics which go along with it abandon us. This is evident when we see the problems of biodiversity and sustainability, in particular. We behold how the loss of biodiversity is taking on shocking proportions; in one generation the number of species has been cut in half. Is it not possible that this is due to the technical view of reality which characterizes our age?

Sustainability is an ideal means to provide for the needs of our own generation, world-wide, without threatening the provisions for future generations. Why is sustainability under pressure?

The dominating technical worldview dominates by means of a model of management and control—the current economy by which one-sided growth is inherent from the beginning. Growth is a must! Growth at all costs!

Sustainability for the whole of humanity and for earth as a whole is, in this scenario, clearly unattainable.

Granted, we have taken positive steps in addressing the problems of the environment through environmental technical applications, but, because this always takes place within the framework of a technical economy, these steps are in most cases nullified by the policies which follow. We demand a reduction of auto emissions, but then we build new roads like crazy and raise the speed limit to 130 kilometers per hour. We continue to be exploiters, polluters, and looters.

The current, popular perspective on the future remains geared to an expectation of salvation for humanity through technology. The result is that genuine reflection about the purpose of life is avoided and reality is reduced to that which is manageable and controllable. The leading principle is the image of a technical construction, continually gaining strength and influence, within which the reality around us does not have an essential value, but only an instrumental one. In this way, all the plants and animals of earth are seen in the light of the material use which they have for human beings in science and technical applications. Even human beings are increasingly seen as objects to be manipulated and improved.

Werner Heisenberg sketched an impressive image of this situation; "With the apparently unlimited expansion of material power, humanity arrives at the situation of a captain whose ship is so strongly built of steel and iron that the magnetic needle of its compass only reacts to the iron mass of the ship and no longer points north. With a ship such as this, it is no longer possible to determine the proper course."[6]

Our culture in the West finds itself abandoned at sea without a compass. Without a doubt, technical power has increased, but disasters have also increased, dramatically. Technical progress as such is a power which easily turns against its master, humanity, and earth itself. These threats remain mainly hidden under the visible, apparently desirable power of technical effectiveness and economic efficiency. The ethical reductionism of this whole process is scarcely recognized.

---

6   Werner Heisenberg, *Das Naturbild der heutigen Physik*, p. 22, referred to by Hugo Staudinger, Geschichte kritischen Denkens, Christiana-Verlag, Stein an Rhein, 2000, p. 181-192.

Intermezzo: the "empirical turn" and postmodernism

It seems to me a good idea, at this point, to take a brief excursion and look at the recent development of the so-called empirical turn in the philosophy of technology.[7] This "turn" opposes, correctly, philosophers such as Heidegger and Ellul who think that the development of technical applications is an autonomous process. According to them, human beings make their own contributions to the existing developments, but they have, in fact, very little to say about it. Heidegger and Ellul confirm, thereby, the existing developments are inevitable.

The thinkers of the empirical turn are right in resisting this idea. They concentrate on special problems in the practice of technical applications, the so-called "cases" which they abstract from the whole of technical developments. They do not devote much attention to the structural development of technical applications or placing technology within the whole of reality.

The *variety of technical applications* certainly needs attention. That is where we can agree with these thinkers. The ethical problems in this field do not have the same urgency everywhere. In the Reformational Philosophical approach to technical applications, justice is always attempted to be done to a variety of things. The idea of an autonomy of technical applications as a massive and unconquerable phenomenon within which there is no place for human responsibility, as Heidegger and Ellul seem to think, is rejected.

As well, it is proper to emphasize that technical applications take place in a historical, cultural, social, economic, and political context, and that different groups are active with different interests and aims. Each group has a particular influence on the development of technology, but none of them can separate themselves from the continuity of this development itself.

Furthermore, all involved have been formed by the same ideological, historical background, which is an important cause of the ethical problems faced. The philosophers of the empirical turn are wrong when they don't take this issue seriously. Investigating individual "cases" in order to make "rules of thumb" for the future is inadequate. These philosophers seldom attempt to discover the roots of the problem and the coherence between the problems. By fighting the symptoms, the problems are merely shifted and there is no attention given to the common basis of all the problems. Ethically speaking, these philosophers remain, with their proposed pragmatic solutions, stuck in the multiform and endless

---

7   See Peter Kroes and Anthonie Meijers (eds.), *The Empirical Turn in the Philosophy of Technology*, Vol. 20. Research in Philosophy of Technology, Jai Press, Amsterdam e.a., 2001.

"labyrinth of technology."[8]

Only when we perceive technical development with all its variety as one phenomenon, do all the individual practical examples make sense. What is needed is justice to both the general ground structure of technology as well the individuality of specific technical phenomena. To orient oneself to only one of these poles is to not do justice to the other pole. Empiricists, because they one-sidedly give attention to discrete problems (the individual trees), do not see the movement in technical developments as a whole (the forest). A mere analysis of the influence of the "actors" who have an influence on technical developments is to neglect the deeper common motives which are at work. Superficiality and a lack of cohesion in the analysis is the result.

Depth of insight into what is happening in specific "cases" is achieved by seeing them in their connection to each other. In a philosophical, ethical reflection, it is not in the first place a matter of an ethics of certain discrete, technical phenomena, but involves an inclusive, ethical approach via cultural images, ethos, motives, values and norms on the basis of which certain technical applications can be evaluated. The pragmatic turn doesn't do this. The increasing, fragmented specialization of our time encourages fragmented thinking. Unfortunately, we're coming to know more and more about less and less.

Merely giving attention to specific technical phenomena can distract us from what is central, namely, the ethical question concerning a possible reorientation and mentality reversal of our culture. The latter can only become a reality when the motives, aims, values and norms involved are disclosed on a larger scale. Or, to say it with a variation on a well-known slogan, "Think universally (globally), but act individually (locally)."

Postmodernism

The so-called "empirical turn" meshed nicely with postmodernism. Postmodernism arose out of opposition to the great Enlightenment narratives and in reaction to the problems of modern technology. That's why it is said that postmodernism is technologically pessimistic, but it still betrays its origins in the Enlightenment. In this, it is hypermodern.

Postmodernism puts the accent more on one pole of the dialectic of the Enlightenment. Over against the management and control pole, it chooses for the freedom pole for individuality instead of universality. Therefore, it displays more fragmentation and discontinuity than coherence and continuity. Thus, postmodernism shares with the philosophy of the "empirical turn" a lack of a coherent vision of reality and promotes an ethical relativism. As the postmodernist Lyotard teaches us, attention is

8   See Willem H. Vandenburg, *The Labyrinth of Technology*, Toronto, 2000.

shifted from "great narratives" to narrow paths of action.

Our conclusion is that empiricism and postmodernism, with their attention to specific technical applications, provide only a limited contribution to a more comprehensive ethics of technology. In a certain sense, empiricism and postmodernism leave the dominant pragmatic, "case study" technological ethics, as I have described it, undisturbed.

"Case study" approach inadequate

"Case study" technological ethics fails to see the big picture. We could specify many examples of how this is so. I limit myself to one, taken from the Dutch landscape.

With the help of environmental technology, it is possible, at the moment, to raise crops in greenhouses, a widespread industry in The Netherlands, in an environmentally cleaner way. Farmers agree that this is a good idea. At the micro-level, thus in a particular greenhouse, you could say that, compared to the past, progress is evident. But making things greener at one site demands an expansion of production to pay for the extra costs involved. This means an increase in the use of energy, which means more fossil fuels are burned, which increases the air pollution. The environmental situation is thus worse than it was before. That happens when a vision concerning the ethics of technology is limited to considerations regarding individual technical matters alone.

A cosmological and ethical deficiency

We return to the main topic of this chapter! I said earlier that the current popular world-and-life-view in the West is nourished by the spirit and ideology of the Enlightenment. Now it is unmistakably so that we can give thanks to the Enlightenment for a lot of good things. But a lot of evil has come out of it as well.

I would say that the current popular view of technical applications suffers from a debilitating cosmological and ethical deficiency. What is said about the cosmos is of limited value because justice is not done to the many-sidedness, depth, coherence, and fullness of reality. Reality is reduced to that which is scientifically and technically controllable. This is a positivistic cosmology or a techno-cosmology. And no attention whatever is given to the dependence upon and the orientation to the cosmos's divine origin—the transcendental orientation of all things.

Next to this cosmological deficit, we have to do with an ethical deficit. Reality surrounding humanity is regarded as being able to be manipulated as a series of useful objects. This ethical deficit is perhaps best characterized as a lack of love. For example, animals have more and more become "producers" of functions determined technically by human beings. The ideas about therapeutic and reproductive cloning of humans, too, fit into this technical worldview.

The German philosopher Sloterdijk is of the opinion that the influence of the Enlightenment in forming humanity has not gone far enough and actually can never go far enough. Genetically modifying human beings is the next step. Sloterdijk says, let's make the most of it!

Enlightening the Enlightenment

Around us we see the spirit of modernity coinciding with uninhibited technical development. By a vast majority, people in the West still adhere to the motif of the old Enlightenment, the "Aufklärung." However, if we look under the surface, it is becoming more and more clear that our Western culture cannot maintain the old Enlightenment commitments to a belief in absolute freedom and to a belief in an absolute power of technical management and control.

The question as to what "Aufklärung" really means was answered, classically, by the great philosopher Immanuel Kant. Kant said that the emancipated humanity of the Enlightenment does not accept any authority above it. "Dare to use your own reason." Kant's imperative is not just a call to a growth in scientific knowledge or to liberation by a spontaneous act of the will, but a call to the brave decision to manage and control all of human life through scientific knowledge. Reason is accepted as an instrument of power. That is, human beings want to recreate the world through science and technology in accordance with their desires. However, this aspiration of the Enlightenment closes its eyes to the disasters which have followed in its wake.

This is admitted by many. In all sorts of present-day philosophy of culture there is criticism of the evident shortcomings of the "Aufklärung." People have come to see that the concept of instrumental reason has had and continues to have destructive effects. Technology is no longer liberating, but has taken on the form of a power above humanity and above Nature, which, as such, cruelly binds human beings, destroys Nature, and threatens the earth.

In spite of this criticism, in practice most people in the West refuse to abandon their starting point. Some seek to provide a corrective supplement to what "Enlightenment" means. For example, Adorno and Horkheimer desire a further elucidation and deepening of the Enlightenment. Others, such as Hastedt, want to carry on a broad expansion of the project of the "Aufklärung." They say that the program of the "Aufklärung" demands a new ecological ethics and an ethics of giving a more adequate direction to the system of technology.

However they try to make adjustments, such thinkers do not in any way abandon the idea of the autonomy of scientific, technical humanity. The latter is indispensable, as it were. They want to broaden things by reason to a fuller, or broader, multifaceted rationality, which covers more areas. Thus the technical worldview is modified here and there, but retains belief in human autonomy. Even when people plead for a second Enlightenment, whereby there is more attention for metaphysical or spiritual questions, which, they say, in the course of time have been neglected, they still remain loyal to the starting point of the Enlightenment—human autonomy.

With the Reformational Philosophy we stand in a tradition of fundamental criticism of this philosophy of life of the Enlightenment. We see here a continuing claim for human autonomy which betrays an intrinsic, intellectual hubris and arrogant will to power. Although we cannot return to before the Enlightenment and become pre-modern, as it were, we ought to acknowledge this vision's destructive effects and ethically look very carefully at the results.

Can the cosmological and ethical deficits of the Enlightenment be erased? To do that we need another approach that is different from current alternatives. With Rohrmoser I would plead for an "enlightenment of the 'Enlightenment.'" Or, to say this with a biblical psalm, "In Your light do we see light" (Ps. 36:10). The "Aufklärung" must itself be "enlightened" by the divine light of Revelation. We don't have to choose between a technical paradise and a technical hell. We may follow another path which truly transcends that dilemma. And, if I may refer to the metaphor of the ship used by Heisenberg, if the captain wants to get the ship back onto the proper course, then he will have to look up and orient himself to the stars of heaven. Thus "technical culture" must be evaluated from an eternal perspective outside technology itself.

First of all, we must come to recognize that we live in a created reality. Further, in this reality a break has occurred between God and humanity. But by God's grace, restoration in and through Christ has now been made possible. Now we may live with the liberating perspective of a Kingdom of love and peace and a Kingdom within which Nature and human life will, ultimately, be eternally filled with the glory of God. Such a deep, religious rebirth and reorientation casts a bright, hope-giving new light on the ethics of technology.

Erasing the cosmological and ethical deficit

Operating from the acknowledgment that reality is a *created* reality, we believe that the cosmological and ethical deficits inherent to a reductionist, scientific approach to reality cannot be compensated for by an expanded, scientific approach. For example, system thinking can be presented as a holistic approach and can be appreciated for its merits, but it still remains an abstract, scientific approach stemming from an anthropocentric orientation. Humanity may no longer be seen as "Lord and Master," but humanity is still regarded as the self-sufficient "pilot" who finally has the last decisive word, operating from his own internal resources.

An increase of dimensions here, thus a more *"comprehensive holistic approach"* is needed. The *whole* of reality must be acknowledged as a *created reality*, in all ways dependent on God as the Origin. A cosmology which proceeds from this conviction erases the cosmological deficit we have pointed to, just as it erases the moral crisis which is the result of the dominating influence of the technical worldview.

God's intimate involvement with created reality is characterized by His love. By accepting the reality of that divine, loving involvement in His creation and His call to love, the ethical deficit we have spoken of can be addressed and replaced by a different attitude and different actions. The command to love God and our neighbor forms the concentration point of all motives, commandments, values, and norms for humanity. "All the Law and the prophets" depend on this call to love, Jesus said (Matt. 22:40). This twofold expression of love ought to be the starting point for an ethics of technical applications. Love is to be expressed, especially to that in creation which is weak and vulnerable. Loves enables us to discern that everything which has been created is charged with divine secrets. To assent from the heart to this truth means that, next to giving attention to values and norms for technical applications, we are called to ecological and social values.

God calls us to responsibility in caring for the world He has placed us in.

A new vision of culture

As we begin to approach science and technology (and the economy) from this new commitment, I believe that the philosopher of culture Hans Jonas can help us. Imagine, he says, that we find ourselves on the moon. We gaze in wonder at the endless cosmos around us. We can see earth against the stars and are impressed by how unique our small home planet seems to be—the only green planet in our solar system—in this overwhelming setting. We know that on earth an abundance of life is found in a tremendous variety. If we want to survive on the moon, earth will be the source for everything we need to keep on living. But standing on the moon, Jonas says, we suddenly realize that the planet earth is in mortal danger; all of life is threatened by the results of the existing technical and economic developments. It's an emergency! What can be done?

As we ponder this emergency, we begin to see what is needed. A responsible rescue effort to save the earth calls for radical new values. Humanity must begin to value the earth as a Garden to be tended (Genesis 2). Human society must become a "dwelling place of fellowship" and a communal home in which Nature, technology, and culture are seen as being completely dependent on each other and where all living beings have a meaningful place and role to fulfill. While recognizing that any romantic idea of a utopian future must be excluded from the outset, nevertheless, through operating from God's call to loving stewardship, we see what He wants earth to be like.

Every human activity, including technology, ought to begin with tender care for and respectful involvement with our fellow inhabitants of this Garden. Each created thing's inherent nature is to be respected, otherwise life will ultimately cease. This is no idolatry of Nature, making Nature itself an end in itself. To the contrary, it is an acknowledgment of the loving, creative energy of the Creator, which calls us as human beings to respond in tune with Him. We have a mandate to dwell responsibly in the Garden, which is earth, and to support and strengthen every living creature.

This vision of earth as a Garden can be a powerful one if it inspires us to action. It is a metaphor, but more than that, it is an image which can capture our imagination and fill our dreams. God calls us to till the Garden with the ultimate intention of creating a communal dwelling place for humanity. The "dwelling place of fellowship in the Garden" idea entails solidarity with and the dependence of humanity upon the entire creation.

We must come to realize that created reality is given to us as a gift. Human beings are given the opportunity to *disclose* and *develop* that which is inherent in creation, like gently removing the wrapping-paper from a precious gift.

This image of our relation to creation is clearly in harmony with the original meaning of the Greek word *oikomonos*, "one who manages a household"—the root of the English word "economy." Everything depends on what we mean by "managing." In this Garden, where we are building our communal home, caring for, cherishing, protecting, and preserving go hand in hand with cultivating, harvesting, and producing. Our modern focus on increasing the scale of production and accelerating cultural development by technology needs fixing. No longer are we to be the technocratic, controlling, dominant, self-serving Lords of this Garden, but rather wise and compassionate stewards, truly "cultivating" this Garden at a scale and tempo which benefits this symbiotic living of human beings and creation.

As the economist Herman Daly of the World Bank has expressed it, we must maintain, and if possible, improve the potential of fruitfulness of the earth. The fruitfulness of the earth ought to benefit all people, now and in the future. Responsible development of culture is to consciously *live from the interest of the "capital" which has been given to us* but does not allow this capital itself to be depleted or even completely used up. This is the central notion which fits the idea of humanity being a *good steward* of creation.

This new vision of culture gives us an impetus to a fundamental reorientation regarding the technical-economic order. It makes room for growth, but this growth is more proportional and selective than what we currently see taking place. We will seek to conserve the diversity of life forms in the plant and animal world. Ecology, technology, and the economy remain in balance with each other if the natural cycles are not cut off and the springs of blessings do not run dry. Our earth, our Garden city, is to be cherished.

A bracing realism is needed at this juncture. We must recognize that humanity's condition has been radically changed through the Fall into sin and the deep alienation between God and humanity. Thorns, thistles, and death sadly accompany us as human beings and the consequences of it are to be seen and felt in technology, too.

However, through God's redemptive grace in Christ there is now a hopeful perspective for our broken and darkened creation. The real purpose of our existence now beckons—the Kingdom of God. This acknowledgment implies an ongoing struggle in all of creation, including human beings. But there is hope. To expect God's Kingdom makes a huge difference, so different from the materialistic and hedonistic attitude of our age. Jesus said, "For what will it profit a man if he gains the whole world and forfeits his life, his soul?'" (Matt. 16:26, ESV modified)

Traditional humanistic ethical approaches

The question we now want to answer is, which ethical approach is the most suitable for erasing the ethical deficit present in the current ethics of technology?

Two still influential Western humanistic ethical approaches—an ethics of duty (deontology, Kant) and a teleological approach (consequentialism, Mill)—have shown themselves, in my opinion, inadequate to address the complex phenomenon of modern technology. Technical applications are no longer characterized by a simple relationship of humans to tools, but have become a much more dynamic system with the increasing application of science to technology, and with world-wide effects. Modern technology is furthermore entwined with economic enterprises. It's a new game, which neither Kant nor Mill help us to solve.

What are the shortcomings of these two traditional ethical systems?

Duty ethics (Kant) has resulted, through its historic development, in a more pragmatic or even ultra-pragmatic ethics, whereby norms which were once seen as permanent became relativized. But pragmatism has been shown to be a weak way of dealing with technological challenges. Look around you.

Teleological, purpose-oriented, consequential ethics (Mill)is also insufficient. This is so because evaluating *aims* (maximizing benefits for the most people) cannot be isolated from evaluating *means* (what are *good ways* to reach this result?). It is evident that when aims are being reached, attention needs to be given to the path taken an*d the manner of taking that path.* Again, look around you and see if this humanistic approach has led to an adequate way of dealing with the technological revolution. We're in a mess and we need help.

That's why ethicists have now introduced an ethics of rules-of-the-game, which actually is not that much different from a pragmatic ethics.

So, in my opinion, we need a new approach for our technological age. The disaster of the current reign of what I have called "technical ethics" calls for an alternative.

An alternative ethics

In my view, a *"response"* oriented *responsibility ethics* is a superior approach to formulating an ethics of technology in our moment of history. This is an ethics in which *ethos, motives, values, and norms are subjects for discussion that is grounded in a vision of reality which is more than a materialistic, human-oriented one.* We are called to give a "response" to God's call to loving of stewardship of our earth.

This responsibility ethics does not have, as is often thought, a philosophical origin, but rather a *theological one.* Far before Jonas wrote about a responsibility ethics, the theologians Niebuhr and Barth were writing about it. At the founding of the World Council of Churches in 1948, this ethics was already the guiding principle in the discussions about society.

The word *"responsibility"* is an appropriate term in our technological age, for it expresses the fact that everyone who is involved in technological development is called to act as someone who has been given plenipotentiary powers and as an *official guardian or steward.* We do not invent our own standards or try to "pull ourselves up by our own bootstraps," but instead we reject seeing ourselves as autonomous beings in a silent universe, created to *respond* to God's call.

Everyone involved must be able to *give an account* of his or her activities to God Himself. Everyone must clearly be able to indicate on what basis he or she is active. Which primary vision of culture is held to? Which ethos, motives, values, principles, and norms underlie his or her activity?

In the discussions about problematic technological developments, bringing in "ethics" is generally associated with indicating "that which cannot be tolerated," a negative association.

In contrast, within the framework of a "responsibility ethics" we begin with emphasizing that which is positive. With the advent of modern technology, a new range of possibilities has arisen, for example, of helping people in their suffering and needs. In that sense, the idea of a "calling" to service fits in well with an ethics of responsibility in the world of technology. For the world of technology is part of God's world.

A good starting point for a responsibility ethics in our time is to ask the technological "players"[9] involved to become more conscious of the positive potential of their actions, emphasizing their full responsibility. We ask them to be able to *give a public account of their activity*.

Operating from a vision of the earth as a God-given Garden city that is moving in the direction of a harmonious, world-wide "dwelling place of fellowship," a prime concern is to help make the world livable and sustainable. We are all called to help provide the basic necessities of life for all and to help relieve the burdens and suffering of all people, even all of life on the planet. Everyone involved in technology—scientists, developers, users—is called to such a vision and such activity. It is our common responsibility *coram Deo*, in God's awesome presence.

The renewal of motives

We have seen how the current, reigning ethos of technology, and thereby of the sectors which are deeply influenced by technical applications, such as agriculture, the economy, politics, etc., can be described as an aspiration to absolute power and a completely controlling management. This ethos differentiates itself in science as the motive of "knowledge is power" and in technical applications as the motive of "technology for the sake of technology," and thus as the motive of technical perfection "that which can be made, ought to be made." In the area of industrial agriculture, harvesting with the help of unbridled scientific-technical power is resulting in exploitation and predatory practices. Due to a materialistic economy, in which the only values are the power of money and materialistic profit, and due to materialistic politics, the reigning cultural powers are strengthened as they group themselves together as a movement. This convergence of the cultural powers has shown itself to have a culturally dislocating effect. It is an illusion to think that these powers could serve values other than those of the growth of power, increasing scale, and concentration.

In the perspective of the image of the earth as a Garden moving in the direction of a "dwelling place of fellowship," we as human beings are seen as being called, in our cultural activities, to deny ourselves, and to love God

---

9 This is not to say that "responsibility" has the same meaning and form for everyone. Different forms of responsibility are substantial, functional, collective, individual, and professional responsibility. See also: J.O. Koenen, *Ethics and Technology* (TU Delft: 2001) p. 12-22.

and our neighbors. Through such a healthy ethos, the underlying motives for the various activities of culture receive another content than that which is currently popular. Instead of an ethos of power in which human beings operate on the basis of self-interest, the ethos of love points in another direction. This implies that, regarding science, we are called to grow in wisdom; regarding technical applications, we are called to build and preserve; regarding agriculture, we are called to harvest, protect, and care for creation; regarding the economy, we are called to be good stewards; and in politics, we are called to serve and to further the rule of law and public justice. This pattern of divergence from the dominant powers of our age is the real path to follow in seeking a significant blossoming of culture.

By way of elucidation, I would like to say something about these authentic motives for science and technical applications.

Science: growing in wisdom

Modern science is heavily influenced by the mentality of technical management and control. It is not a technical activity due to the results of its applications, but it is technical because it tends only to look at reality through the frame of measurability and predictability and because it is, in general, only interested in managing and controlling reality. Let there be no misunderstanding. Many scientists will say that they want to come to know reality better because they are curious or because they are interested in the truth. However, this does not alter my assertion that modern science has become technical in its core being due to the supremely dominating cultural motives and cultural powers driving it.

Science can only relate properly to the fullness of reality when an acknowledgment of the origin and the meaning of reality precede it. Science ought to be integrated into the full reality of experience and deepened as a form of the knowledge of reality. Then, scientific knowledge will serve our growth in wisdom.

This method of science will enhance an increasing, comprehensive insight into reality. For science's aim ought to be to gain insight and to strengthen human responsibility vis a vis all that is happening in reality, conceiving that reality as a Garden grows into a "dwelling place of fellowship."

Seen in this way, an interdisciplinary approach in science is very desirable. We are far too little aware that technical applications need a more comprehensive basis and that that basis, as a contribution to growing in wisdom, will lead to a creative and careful way of being active in the realm of technology and serve life in all its dimensions. If a more interdisciplinary approach had been supported with regard to technology, then biology and ecology, for example, would have been accepted as basic sciences for technical science a long time ago.

Technology as prosthesis

What I have said about science in general, also applies to technical science and to technology, that is, in my terms, the study of technical applications. Technology too ought not to be used as a mere instrument of scientific-technical domination and management.

Instrumental conceptions of technology do not do justice to the place of surprising discoveries in technical development. Many have said that discovery is the heart of technical applications. That is, human creativity ought not to allow science to restrain it. Instead, creativity ought to be stimulated by old and new scientific knowledge. In the university study of engineering, there ought to be more attention given to genuine creativity in discoveries and innovations in order to make technology more geared to serving life. For, as we have noted, the motive of technical applications ought to be to serve life and society. Technical applications should function both individually and collectively as a prosthesis, as it were. In that case, human beings remain in charge of technology instead of being its slaves.

And here we don't have to think only of small-scale technical applications. In the case of The Netherlands, an example of a large-scale technical application is the dam across the Oosterschelde, a dam which can operate half-open to the ocean. This is a prime example of deviating from existing trends and seeking to preserve safety, the environment, and Nature.

"And what about the costs, then?" I hear as a possible objection. The costs will indeed be higher with these ideals. In general, modern technical applications are, in fact, much too cheap. This is so because we only take account of the initial economic, or, more accurately, the production costs while very often not paying attention to the long-term damage inflicted upon Nature and the environment. Or, to mention another issue, safety is sacrificed for efficiency. These are consequences of a blind acceptance of the technical worldview as a guide.

It is especially with modern technology that we see clearly how we are not really focusing sufficiently on being of service to the earth and to society. With the increase in the complexity of the technical applications, a reckless, nonchalant attitude has increased. We would do better to offer more resistance to technology's temptations. Humility is called for, instead of arrogance, in our life here on God's earth.

Other values

What are ecological values? Here are some—preserving biodiversity, striving to have clean water, soil, and air, maintaining and creating fertile soil, and generally improving the environment in which we live. The biosphere must be left intact, and therefore we have to combat dangerous gas emissions. Technical applications have to be adapted to natural life environments, and the diversity of life should be maintained.

The technical and economic values regarding livability, safety, trustworthiness, satisfying basic needs of food and health, the struggle against suffering and disease, and fighting threats from Nature are all aimed at healing, sustainability, diminishing physical burdens at work, etc. Beyond our physical needs, true purpose in life is found in spiritual growth, personal relations, and life in community. Thereby we see how technical values also are connected to social values.

These social values are those of the spirit of community, of frugality, of justice, of caring, of strengthening information and communication, and thus of social well-being in general. Is it too audacious to say that "rest," "having free time," and "spiritual flourishing" may also be mentioned here as the forgotten social values of our technological age?

The integral framework of norms

With relation to technical developments, I have up to this point discussed our image of culture and the ethos, motives, and the values which we must constantly have in view. Technology must serve the great diversity of the existing forms of life and be suited to a responsible development of the "Garden" we live in. Our context is always the existing situation.

We are called to travel in a "normed" direction. In order to see whether or not we are going in the right direction, numerous principles or values, with the concrete norms flowing from them, must be our touchstones. These principles have to do not only with technical applications but also have to do with the variegated relationships which these technical applications have with human life, Nature, and society.

The integral framework of principles or norms derived from the philosophical concept of structures, that is, the cosmology of Reformational Philosophy, forms a helpful grid for arriving at responsible technical development. These norms are: the cultural-historical norms, the norm of effectivity, the norms of *harmony* between continuity and discontinuity, between large-scale and small-scale, between integration and differentiation, between universality and individuality, the norms of clear information and open communication (with all the "players" involved), the norm of harmony between human beings, technology, Nature, and society, the norms of stewardship, of efficiency, of doing universal justice to the "players" involved and to Nature and culture, the norms of care and respect for everything and everyone involved in technical development, and the norms of service, trust, and faith. Within this framework we must distance ourselves from the thought that we can take irresponsible risks that put safety in jeopardy.

In my publications, I have given attention to this integral framework of norms.[10] Honoring such a normative framework helps us to replace one-sided technical development with a responsible, richly varied unfolding of Nature and society. Technology may not be allowed to suppress Nature and society, but ought to serve them. Our aim must not be a one-sided or one-dimensional technical culture, but rather a rich blooming of culture.

Consequences of a reorientation in culture

I would now like to go on and discuss the consequences of a new cultural approach. It is important to remind ourselves that I have painted a picture of the two different orientations or perspectives in a somewhat artificial, black-and-white fashion, over against each other. In reality such an absolute opposition does not exist.

When we acknowledge the existence of a created reality, we see better what is happening. The technical worldview is a kind of parasite that lives off created reality. It disturbs that reality, but can never operate separately from it. That is also the reason why within a technical worldview one can observe so much flexibility, but also that, more often than not, that kind of a worldview proves insufficient in practice for many people. Happily, however, there are inconsistencies here which can be very instructive and helpful. Conversely, we who acknowledge reality as created are too often unconsciously, unhealthily bound to a technical worldview, something that ought to cause us pain.

In connection with this, it is significant to follow the discussions about the cause of the destruction of Nature and whether this destruction could have something to do with the cultural mandate of Genesis 1:28: "Subdue the earth..." From within a technical worldview, to appeal to this cultural mandate can lead to big problems since little attention is given to protecting, caring for, and sustaining the earth. In my view, making an appeal to this mandate within the purview of the image of the Garden as "dwelling place of fellowship" will lead to a much more harmonious development of technical applications. That's why I prefer to speak of a *creation mandate* instead of a *cultural mandate*.

In any case, it is very clear that to choose the "good path" is a charge continually given to us all, implying struggle and excluding indolence.

A "guide to the lost"

Up until now I have wanted to approach the phenomenon of technology from another ethical perspective than is currently popular. The question arises as to what this perspective means for the different, indi-

---

10 See E. Schuurman, *Faith and Hope in Technology* (Toronto: Clements, 2003), p. 193-200.

vidual technical phenomena—called "cases" by many. Earlier I said that when "cases" are looked at, at the micro-level, we can see how ethical improvements can be made, but that in relation to the phenomenon of the macro-level, they usually don't lead to much more than temporary solutions.

The ethical perspective, as I have sketched it, with attention to images of culture, ethos, motives, values, and norms, would have to be worked out in a cohesive, ethical evaluating framework. Such a framework could be used as a "checklist" when analyzing and evaluating practical, individual technical matters. Then the "cases" which are given an account of are connected to a responsible development of technical applications in general. In other words, for those who have gotten lost in the technical maze, such an ethical evaluating framework can be used as a good guide. This doesn't mean at all that different actions are always the result, but rather that these actions are *performed differently*; although choosing for another direction can also mean choosing for different technical applications in certain cases.

Priorities

To give an example, it is very common in science and technical applications to strive to achieve extraordinary, headline results. The result is that the more mundane area of simple social justice is neglected. Less attention is given to technical applications which would help many people in the fight against disease. We see this in the case of the exorbitant, money-devouring projects in the area of space travel. Not that such projects are not interesting, but in terms of costs and social justice, shouldn't we have different priorities?

And—to mention another example of injustice—isn't it necessary to share our God-given raw materials so that the poor and the needy who are living in our "dwelling place of fellowship," may also share in them? This rearranging of priorities fulfills the call to provide for everyone. Hunger is caused, in many cases, by a one-sided technical-economic development, "*There is enough for every need, but not for every greed.*"

Shortly we'll see that conscious planning in relation to future technical-economic developments is very important to emphasize in politics. Here we can say that, with regard to the right priorities, attention should be given to proactive reflection-ahead-of-time, instead of the current tendency to reflection-after-the-event.

Adaptive technology

Modern technical applications should connect with the actual, unique situations in which human beings, culture, and the natural landscape exist. As such, modern technical applications ought to be ecologically and culturally responsible, that is, *adapted* technical applications. Where we see dislocation, we must work to restore harmony. How much damage to Nature and the landscape could have been restored or prevented by greater care? This does not mean that we must return to the age of technical applications carried out by traditional craftsmen. The modernization of technology is a given.

Differentiation in technical development will clearly need to have a cultural dimension. Technical applications ought to enrich culture, not undermine it. Sadly, we see the latter in much of the developing countries. Introducing modern technology means, in many cases, a break with the existing culture instead of an adaptation to existing cultural diversity. Sprawling, filthy cities for masses of the poor and that—strangely enough—exist next to often super-modern industries, have a sad consequence; the depopulation of the countryside and thus the destruction of age-old cultures.

But in the developed world with its industries there are serious problems, too, which have to do with over-development, so that Nature suffers under technical applications. We must become more and more aware of the fact—and this applies to the engineer, the technical scientist, and the technician, in the first place—that damaging waste products are a contradiction to responsible technical applications.

It is a shame that the still fertile ideas of Schumacher, with his call for more adapted intermediate or small-scale technologies, are receiving so little attention. There was a lot of attention for them during the energy crisis of the 1970's. When the crisis waned, people often made a caricature of Schumacher's ideas. He did not mean that we should return to primitive and pre-scientific technology. Rather, he encouraged a technology which connects to Nature and culture, and thus suitable to our human size. We possess tremendous technical power, but we remain dependent upon fragile eco-systems. The existing technical-economic powers do not, in general, reckon sufficiently with these systems. That's why creative, inventive technology is called for, with room for discovery and innovation, applications which are economically productive, ecologically and culturally adapted, socially just, and which satisfy personal and communal life. Computer technology and the Internet offer us possibilities here, because through them decentralization of power is feasible. When technology empowers us, then the vision must become even clearer that we must use it with wisdom.

Adapted agriculture

To give an example, this means that there are consequences for industrial agriculture. In industrial agriculture, we have been repeatedly guilty of "technicalizing." On the one hand, modern technology makes it possible to lower production costs, but on the other hand, the increased production often is damaging to the farmer, the animals, and the environment.

Some of the obvious problems are the exhaustion of the soil, the pollution of the ground, new soil diseases, the disturbance of the landscape, the loss of biodiversity, and an uncertain future for the countryside. While in agriculture we are interacting with a living reality, we have too often regarded the latter as a non-organic machine. Possessed by a scientific-technical ideal of control and management, agriculture has been alienated from its ecological, biotic, and cultural context.

By encouraging ecological-organic agriculture, we want to help restore the health of relationships. Qualitatively good products and benefits for the environment can go together. In this kind of agriculture, we are not returning, romantically, to a bygone age, but, with a qualitatively high level of biological knowledge and the soil sciences, we are more wisely interacting with the soil, plants, and animals.

Genetic modification

The lack of a normative framework of new technology is especially evident in the introduction of genetic manipulation of plants, animal, and human beings. In applying biotechnology, in general, justice is not done to what life really is. This happens, for example, if the genetic structure of living things is compared to a structure of Lego blocks. Because of this kind of reductionism or even abandonment of the value of life, it is no wonder that biotechnology (genetic modification) is finding itself wrestling with unforeseen problems.

Genetic modification must be looked at critically and ethically and be legally controlled. The current developments are unpredictable and hazardous, and their negative consequences are possibly irreversible. Another approach, different from the technical model, should be adopted in dealing with living organisms. That new kind of model must see organisms as living wholes. Then life will be protected and cherished.

In general, I would advocate using the "No, unless" principle, as far as the possibilities for genetic manipulation of plants, animals, and human beings are concerned. This is a principle which, happily, has been accepted by many others as well. This "No" is intended to prevent us from causing disease or disturbances to Nature or the loss of biodiversity as a result of genetically manipulated plants. If people still insist on introducing such techniques, then this must be allowed only on the basis of good argumentation. Entrepreneurs alone ought not to bear the sole responsibility of the risks involved in this area. Political decision making with a framework for evaluation is called for.

It is clear that if genetic modification of human beings is being considered, only particular physical organs are in view. Genetic manipulation via the germ cell line, whereby the entire human being is involved, must remain forbidden. The theoretical possibility of therapeutic and reproductive cloning of human beings has, thankfully, till now only been found acceptable by a small minority. Let's hope this remains the case!

Alternative energy

The whole question of alternative energy is pressing. We need to become more creative in developing alternatives to unsustainable, climate-changing, and polluting forms of energy. In the meantime, we can do more to maximize clean energy from existing sources, for example, from waste incineration plants. And we should promote recycling as much as possible in both short and long term.

Already a great new wave developing new, natural energy sources is sweeping the world. All around us, new technologies based on abundant, inexhaustible, and/or renewable energy are appearing, such as hydrogen, photocells, biomass, wind energy, thermal energy, and tidal flows. We must welcome the opportunities while being very conscious of the dangers of a merely technical approach to the problems of energy.

Dilemmas

Dilemmas can multiply themselves in this attempt at a reorientation of technology. How are transitions from the old to the new forms of energy to be carried out? Existing technology cannot be totally replaced in a twinkling of an eye. There are urgent problems to be solved regarding these existing technologies. One is the problem of radioactive waste materials used in nuclear reactors. We must carry on research in order to make such radioactive materials less harmful. If the latter is not possible, a complete eradication of nuclear energy may be the only alternative in the long run.

Political action

We are becoming more and more aware of how many human activities are deeply entwined with each other. Within our "free market economy," if we conceive of "freedom" as "freedom with accompanying responsibility," something like the ethical framework which I have sketched will be a growing necessity.

Reality teaches us, however, that economic powers strengthen "technicalism," rather than weaken it. The political arena is one central place where damaging, wrong moves can be exposed and resisted. We must seek to put a stop to the "technicalizing" of society through political action. We can certainly begin to choose another direction for technology, in favor of broadly normed and differently structured technology; a direction which is more friendly to the environment, to animals, to human beings, and to culture. New juridical processes will demand that people act in a more consciously ethical way in the lab, in industry, and in society as a whole.

We must acknowledge that such a new national political move (I'm thinking concretely of The Netherlands here) can only be effective if—in the light of our globalized culture—this new direction is supported in the international political arena and in the light of mutual agreements, international law, and world-wide public justice. The prophetic message of Amos is relevant for the whole world, "But let justice roll like waters, and righteousness like an ever-flowing stream." (Amos 5:24)

There are enough examples in history which indicate that politics can straighten out developments which have grown crooked. We see enough examples in history which show us that corrective legislation is possible regarding child labor, social security, housing policy, price policy, environmental protection, and quality and safety standards. Governments have created frameworks for responsible entrepreneurship and responsible technology. We must strive to further the orientation of technology in the direction of a truly *serving* mentality.

Accountability

When new developments begin to appear, it is correct to speak about the principle of taking steps sufficiently ahead of time to assure that things go well. That means that democratic governments make a special effort to prevent new developments from causing harm. Can those in government really have an complete overview of all the possible consequences of new technical-economic developments? Of course not. Nevertheless, the government can make a positive contribution in this area.

Let's look at one example. A federal government can hold technology business concerns accountable for the harmful effects of their activities and products. Up till now, the costs of dealing with such results have been generally borne by the federal government, and thus by the tax payer. This can be arranged differently. Technology businesses can be held liable for the social and environmental damage they cause. The right to entrepreneurship has its own flip side, that is, accountability.

Struggle and Hope

Finally, I hope that I have not left the reader with the impression that we can easily achieve the new orientation and direction I have sketched. I do hope that my thoughts may have a stimulating, positive effect.

According to the Bible, "thorns and thistles" and "sweat and tears" will always accompany all our work, whatever it may be (Gen. 3:18,19). This will be so until God finally intervenes and the earth as it is now, deeply disturbed in its development, will be miraculously transformed

into the divine Garden city (Revelation 21:9-22:5), in which human beings will be revealed as liberated people, liberated to obtain the freedom
of the glory of children of God (Rom. 8:21). In a totally surprising way
,it will then appear that the work performed in science and technology,
in many cases in spite of the scientists and technicians themselves, is intimately connected to God's new creation.

That perspective gives hope and creates responsibilities. That hope
and those responsibilities inspire an aspiration to a new kind of ethics, one
in the light of which people are asked to take on their responsibility to
seek the purpose of technology not apart from, but as a part of the whole
purpose of reality—the Kingdom of God. That ethical perspective, as I've
tried to sketch it, continues to be inspiring. That kind of ethics must also
be proclaimed and given a hearing in the technological world of our day.

Literature

Barbour, Ian, 1993, *Ethics in an Age of Technology*, San Francisco: Harper

Davidse, Jan,1999, *Het is vol wonderen om u heen—Gedachten over techniek,
cultuur en religie*, Zoetermeer: Meinema.

Doorman, Joop, *Onderwijs in Techniek en Ethiek*, Bestuurscommissie Techniek
en Ethiek, Technische Universiteit Eindhoven.

Durbin, Paul (ed.), 1987, *Technology and Responsibility—Society for Philosophy
and Technology*, Dordrecht: Reidel.

Hastedt, Heiner, 1991, *Aufklärung und Technik—Grundprobleme einer Ethik der
Technik*, Frankfurt am Main: Suhrkamp.

Howe, Günther, 1971, *Gott und die Technik. Die Verantwortung der Christenheit
für die technisch-wissenschaftliche Welt.* Hamburg: Furche.

Houston, Graham, 1998, *Virtual Morality—Christian Ethics in the Computer
Age*, Leicester: Apollos/IVP.

Jochemsen, Henk (ed.), 2000, *Toetsen en Begrenzen—Een ethische en politieke
beoordeling van de modern biotechnologie.* ChristenUnie, Amsterdam:
Buijten & Schipperheijn.

Jonas, Hans, 1984, *The Imperative of Responsibility. In Search of an Ethics for the
Technological Age*, Chicago: Chicago Press.

KNAW, 1994, *De Forumrol van de KNAW inzake ethische aspecten van
wetenschappelijk onderzoek.* Commissie Wetenschap en Ethiek,
Amsterdam.

Kranzberg, Melvin (ed.), 1980, *Ethics in an Age of Pervasive Technology*, Colorado:
Westview Press/Boulder.

Kroes, Peter, 1998, *Zin en onzin van ethiekonderwijs voor ingenieurs*, Lecture
at Delft Technical University.

Kroesen, J.O., 2001, *Ethics and Technology*, lectures at Delft University of
Technology.

Mead, Margaret, and Polanyi, Michael, et al, 1986, *Christians in a Technological Era*, New York: Seabury.

Mitcham, Carl (ed.), 1989, *Ethics and Technology—Research in Philosophy and Technology*, vol. 9, London: Jai Press.

Mitcham, Carl (ed.), 1998, *Technology, Ethics and Culture—Research in Philosophy*, vol. 17, London: Jai Press.

Noble, David, 1997, *The Religion of Technology—The Divinity of Man and the Spirit of Invention*, New York: Alfred Knopf.

Rohrmoser, Günter, 1996, *Landwirtschaft in der Ökologie-und Kulturkrise, Gesellschaft für Kulturwissenschaft*, Bietigheim/Baden.

Schumacher, Ernst F., 1973, *Small is Beautiful*, London: Blond & Briggers.

Schumacher, Ernst F., 1977, *A Guide for the Perplexed*, London: Jonathan Cape, Ltd.

Schuurman, Egbert, 1972, *Technology and the Future—a Philosophical Challenge*, Toronto: Wedge.

Schuurman, Egbert, 1990, *Perspective on Technology and Culture*, Sioux Center, U.S.A.: Dordt Press / Potcheftstroom: Institute for Reformational Studies.

Schuurman, Egbert, 2002, *Faith, Hope and Technology*, England: Piquant Press.

Sloterdijk, Peter, 2000, *Regels voor het Mensenpark*, Meppel: Boom.

Staudinger, Peter, 2000, *Geschichte kritischen Denkens*, Stein am Rhein: Christiana Verlag.

Strijsbos, Sytse, 1998, *Het Technische Wereldbeeld—een wijsgerig onderzoek van het systeemdenken*, Amsterdam: Buijten & Schipperheijn.

Sweet, William, 1998, *Religious Belief and the Influence of Technology*, p. 249-67 in Mitcham (ed.), *Research in Philosophy and Technology*, London: Jai Press.

Tillich, Paul, 1986, *The Spiritual Situation in Our Technical Society*, New York: Scribner.

Vries, Marc, J. de, 2001, *The Postmodern Technological Society: a Critical Perspective,* unpublished lecture, Potchefstroom.

Vandenburg, William H., 2000, *The Labyrinth of Technology*, Toronto: University of Toronto Press.

Wal, Koo van der, 1996, *De Omkering van de Wereld—Achtegronden van de milieucrisis en het zinloosheidsbesef*, Baarn: Ambo.

Wauzinsky, Robert A., 2001, *Discerning Prometheus—The Cry for Wisdom in Our Technological Society*, London: Associated University Press.

Zandvoort, Henk, et al., 2000, *Ethiek en Techniek*, syllabus, Delft University of Technology.

CHAPTER 5

THE CHALLENGE OF THE ISLAMIC CRITICISM OF TECHNOLOGY[1] [2]

Introduction

Due to the rise of terrorism in the world, we have become very conscious that the Western world and the Islamic world, while having a common history, also have many differences. There is an increasing atmosphere of mutual confrontation. I would like to look at these two uneasy worlds as they face modern technology and the problems associated with it. This is not a common approach to the tensions between Western and Islamic culture, but—as we shall see—it is certainly a very significant and fascinating one.

When we place this subject in historical perspective, we cannot avoid the religious background of technology in the Islamic world and in the West. Here I find a connection with another burning issue; that of the vitality of religion in the world and the arguments regarding its influence upon culture (Haberman, 2005), in particular upon technological developments.

In relation to the concept of "religion," some clarification is called for. When, for example, the media give attention to religion, they treat it mainly as one of the many factors of human life, next to sports, politics, or science. However, if we look carefully at religious communities and at different social structures in the world, we can see that religion is not only one typical function or variable among others, but is the *root* from which the different branches of life sprout, grow, and are continually dependent. That will become evident in the course of this essay.

---

1   Concluding lecture, September 20, 2007, Wageningen University.

2   Various Muslim and Christian scholars have commented critically on this lecture: *Different Cultures, One World—Dialogue between Christians and Muslims about globalizing technology,* H. Jochemsen & J. van der Stoep (eds.), Amsterdam: Rozenberg, 2010.

After I have given a brief sketch of the history of technology in the Islamic world, I will discuss the backgrounds of the current, growing tensions between Islam and the West. Different Islamic ideologies will be briefly looked at. Science and technology play an important role in their thought. The Islamic criticism of technology stems from two sources; on the one hand from the spiritual, peace-loving Muslims, and on the other hand from the radical, violent branch.

I try to clarify the tensions which exist in this area for the West by examining the internal tensions which are present in the Western culture. It becomes evident that these especially go together with modern technology. These tensions have been present for a long time, but have been strengthened since Christian culture began to be secularized under the influence of the Enlightenment. The latter cultural movement does not want to be linked to religion, but has taken form as a radical movement in a definite and visible way which is more and more at odds with Christianity. When I describe this as a *religion* of a closed, material world—where people are blind to the non-material dimension of reality—I am doing so in order to portray more clearly the total picture of the area of tensions between Islam, Christianity, and the Enlightenment in relation to modern technological developments. This helps us to analyze the problems and formulate solutions to them.

In confrontation with both the Christian-philosophical and the Islamic criticism, Western culture is challenged to self-reflection. In my opinion, a real change to the dominating Western culture paradigm is needed because the ethical framework within which Western culture has developed and is developing. This is important because we have to do not just with an isolated Western culture, but with world-wide problems. The tensions, as well, with certain movements within the Islamic world could hereby possibly be able to be reduced. Islamic terrorists, however, will not be satisfied with this, for their attitude reflects—as they themselves say—a firm religious position which they will never give up. At the most it might be possible to dampen their hostility by seeking to overcome evil with good (Romans 12:21).

Technology and Islam

Let's start by looking at the place of science technology in the Islamic world.

Since the death of Mohammed in 632, history teaches us that early Islam was strongly influenced by the Greek-Hellenistic world. Thereby a favorable climate developed for carrying out and furthering science with a special Islamic tint (Stöklein und Daiber, 1990, 102). Science was seen as taking place in a universe created by Allah. That universe displays order and balance, and is therefore an aesthetic unity. The philosophy and science which developed in that perspective flourished in the Islamic world for more than 500 years. The climax of this development was the civilization of the Arabian world in the 9th and 10th centuries A.D. In this period, we also see much knowledge imported from foreign sources—for example from Persia, India, and even China. That was in accordance with the Islamic mandate that everyone increase one's knowledge one's whole life long. Scientific experimentation and posing technical questions were not at all strange for Islam at that time. Thereby, attention to Nature had to take place in a careful way, just like the care a husband and father ought to give his family. This development furthered trade (the economy), which then had a beneficial effect on science and technology. Historians say that at that time there was a symbiosis between Islamic religion and (practical) science. Building religious houses, mosques, schools, and carrying on water management in desert countries, in particular, were significant illustrations of this.

It is clear that the Islamic peoples were ahead of Western countries in the area of science and technology up till the Middle Ages. At the beginning of the Middle Ages, Islam even functioned as a bridge between the antique Western world of Greece and Rome and contemporary Europe. The West has a lot to owe to the Arab world as far as scientific development is concerned.

However, since the 11th century, the development of the sciences has been stagnant in the Arab states. All kinds of reasons are given—especially political and social-economic—for this decline. Since that time, traditionalism and isolation have been the chief characteristics of the Islamic world. This goes together with a change from a positive to a negative evaluation of science and technology (al-Hassan, 2001, Hoodbhoy, 2007).

Later, during the industrial and post-industrial ages, the Islamic countries contributed little to the development of science and technology, with the exception of the oil-rich Arab countries, which via oil production and industry have helped develop related technology, and also with exception to weapon technology developed in Islamic countries. It is striking that, at this moment in time there are Islamic scholars who are again—as I hope to show—seeking to advance modern science and technology in the light of Islam's own, original history and sources (Zayd, 2006, 31-35). Their criticism is not of science and technology as such, but of the "technological culture" of the West, thus the Western ethics bound up with this technology.

The influence of the Enlightenment in the West

In the course of time, the West, with its dominating faith in progress—especially stimulated by the Enlightenment—began to feed the bias against Islam to the effect that Islam itself was opposed to science and technology. The contemplative character of Islam and the fatalistic attitude toward life of the Arab was blamed for this. Although contrary to its original attitude, this ethos has indeed had much influence in the world of Islam. The opposition to Western science and technology has been strengthened by it. Since the 12th century, the Islamic world has been looking more to the past than to the future.

In the 20th century, a change took place through the process of globalization. Universities in the Arab world were founded as a result. And much was taken over from the West (Huntington, 2001, 70, and Soroush, 2000). But it seems that modern technology has only been valued to the degree it could be used to serve Islamic religion. The idea is that science and technology must be brought under the Islamic flag. That cannot take place easily, however. Historically, Western ethics has accompanied the development of technology. Islamic resistance is being offered to such ethics just as resistance continues to be offered to the faith in progress present in the Western background of those ethics. Islamic acceptance of scientific and technological knowledge—modernization—continues to form a glaring contrast with the opposition to westernization, to secularism, to materialism, and to Western profanity (Soroush, 2000, XVII). The conviction is that modernization must be supplied with a moral compass via Islam (WRR, 2006, 38,39).

Reactions to Islam

It is furthermore important to distinguish between different reactions from within the Islamic community. For more than one Muslim country, various reactions even go back to the colonial period. First we have to do with the radical, violent, fundamentalist stream which rejects both the acceptance of science and technology and westernization (the ethics of the Enlightenment.) Then there is the stream which clearly takes over both elements from the West. This concerns especially the political and economic holders of power. But it is also true of some Muslim scholars (Hoodbhoy, 2007, 55). It is understandable that the first stream also opposes the latter phenomenon, and this is the reason why terrorist actions of radical Muslims takes place just as often in Muslim countries as in Western ones.

We also have to do with Islamic reformers, as Huntington calls them (2001, 118 f.). Others call this group spiritual and peace-loving. They accept modern developments in science and technology, but without the dominating Western ethos. For them, it is a matter of accepting science and technology, whereby there is support for a thorough process of proper (non-secular) "rationalization" going together with spiritual conviction (Hassan, 2007, Soroush, 2004). In the same way they argue often for the approach to society of the western democracies (Soroush, 200, WRR, 2006, 29-58).

As a result of substantial differences regarding convictions and the growing tensions between these three streams, political resistance and violent protest against the West will most likely grow, and within the Western world—where Islam is growing fast—cultural tensions will increase. The choice of the fanatical Islamic stream, the smallest group, is resulting in a violent threat to Western culture. Witnessing this passionate impulse to destruction leaves us with a somber perspective on the world situation.

Enemies of the West

The study of Ian Buruma and Avisah Margalit about Occidentalism (Buruma, 2004) provides us with good insights into this process. That which they call "Occidentalism" is the ugly image of the West which is painted by its enemies. According to this narrative, industrialists, capitalism, and economic laissez-faire policies have spread like diseases from the West, under the leadership of America, and infected the entire world. The fanatic Muslim sees in this "Americanism" a mechanical civilization which ruthlessly destroys cultures. Globalization strengthens the power of this evil civilization, which is cold, rationalistic, and soulless. Granted, the spirit of the West is capable of developing technology, bringing it to a higher level, and realizing great economic successes, but it cannot aspire to the higher things in life because it totally lacks spirituality. It is a hopeless situation since this spirituality is so important, indeed, the most important thing in life. The spirit of the West spreads atheistic scientism, a faith in science and technology as the only way to gather knowledge (Buruma, 2004, 76,96). For Muslims, Western religion is identical to materialism and this religion is thus in conflict with the worship of the divine Spirit.

According to Buruma and Margalit, the roots of this antagonism to the West lie in the opposition to its "technological culture." The spirit of the West is brain-sick, arrogant, superficial, and hostile to piety. Instead, it is efficient, like a mechanical calculator. Western culture is therefore a non-spiritual, materialistic culture of technical, proud presumption that is power hungry, covetous, brutal, and decadent, which deserves to be destroyed. Suicide terrorism, on this basis, forms the pinnacle of hostility to Western technology. The suicide terrorists, as worshippers of the divine Spirit, carry out the slaughter of the worshippers of earthly matter,under the ambiguous motto (I quote): "Death for the sake of Allah is our highest ambition." (Buruma, 2004, 73; Al-Ansari, 2007). Their struggle against the West is a matter of holy dedication to the cause.

Islamic terrorism and the dialectic of Western culture

In their analysis of Occidentalism, Buruma and Margalit do their best to understand the enemies of the West. They say that without understanding why these Islamic movements hate the West so much, we cannot prevent them from being destructive in the extreme. (Buruma, 2004 17)

It is quite easy to see parallels between Buruma and Margalit's analysis of Western culture and their search for reasons for the hostility to this culture, with what we call in the Reformed Philosophy movement the dialectic of Western culture. It is striking how often Buruma and Margalit look for the causes of terrorism in the internal tensions of Western "technological culture" itself. These tensions have been felt world-wide since the beginning of globalization. While the reactions up till a short while ago remained limited to Western culture, recently there are counter movements to be seen across the world, including the Far East. Islamic jihadistic terrorism is the strongest and most dangerous expression of this. Thereby, repeated use is made of Western cultural criticism. Thus Heidegger's criticism of "technological culture" is popular with radical Islamists (WRR, 2006, 45); Zayd, 2006).

What is the dialectic of culture?

What do we mean with the expression the "dialectic of culture"?

My inaugural lecture treated the cultural tension between technocracy and revolution (Schuurman, 1973). Since that time, the subject of our cultural dialectic—tension or conflict—has been a recurring theme in my lectures in various different ways. This cultural dialectic reveals to us what is happening in our culture at the deepest level, what the problems are, how serious they are, and also—when we pay attention to the origin and the historical developments of it—how it should and can be resisted.

Herman Dooyeweerd (1959, 10 f.) saw the origin of the Western dialectic in the pretentions of a humanity which imagines itself to be self-sufficient and autonomous, a humanity without God. As a consequence of this, the world is accepted as an anthropocentric, closed world, and history is regarded as a purely human history. Because in our culture the openness to a transcendent God has been shut down, human beings—in whatever variations conceivable—see themselves as only involved with a "diesseitige," earthly reality.

People of the West attempt to make a reality of the idea of a self-glorifying autonomy in science, and later, as confirmation, in technology. The thought that human beings and the world can be brought to perfection by modern technology has taken over. This development has set powers loose which have raised the tensions in the world to a fever pitch. The ideal of unimaginable material prosperity may partially have been realized but at the same time, it has become clear that this has taken place at the expense of human freedom and our environment, and that we with our prosperity are living on a volcano which is about to explode. Western culture is a culture which is deeply divided against itself. An absolutized freedom becomes antagonistic to the absolutizing of scientific-technical control and management, and vice versa. This tension works itself out in history.

The development of the cultural dialectic

At the beginning this dialectic—which at its base has thus a religious character—was above all a philosophical, theoretical matter, but since the infiltration and eventual hegemony of the Enlightenment has become dominantly cultural in the broadest sense. It is certainly in the spirit of the Enlightenment to not only know reality but also to organize and give form to it rationally. The idea is to develop a society with the instrument of reason, a society in which human freedom would be able to express itself and become what it wants to be. The actual situation, however, is that when imagined, objective structures developed by autonomous reason are erected, they turn into independent powers in their own right, and as such become enemies of cultural freedom. This threat becomes greater in proportion to the dynamic and complex development of these powers, so that human beings do not have oversight any more, let alone being able to change this development.

In my lectures on Modern Philosophical Streams, I have always tried to show how the powers of science, technology, and economy are recommended and strengthened by dominant philosophical streams such as positivism, pragmatism, and system thinking. This is so particularly because these streams consider new developments in technology to be necessary as solutions to the cultural problems which have been brought into being by obsolete technology.

Over against this are those philosophical streams which represent the opposite pole of the dialectic. Thus, we see existentialists demonstrating that human freedom is under threat in technical society, where human beings themselves have been degraded into technically manipulable objects. The neo-Marxists have called us to recognize how the economic and political powers exercise a consolidating influence, strengthening the development of science and technical applications, with the result being a threat to human beings as bearers of culture, a threat to human beings in their political roles. The thinkers of the counter-culture call us to pay attention to the suppression of Nature, and therefore argue for a reduction of technical applications which destroy Nature and pollute the environment. New Age thinkers oppose materialism and aspire to a more spiritual way of life, and philosophers of the Green movement emphasize above all the significance of Nature as a totality, over against a massive, artificial, and abstract technology (Schuurman, 2003, 135-161).

Characteristic of our age is how *all* people feel the tension in our technological culture; you could say that they experience this in both body and soul because the tension is growing greater by the minute between an infinite technical drive to expand itself and the finiteness of the creation and its hidden possibilities.

The primacy of the scientific-technical ideal of control

That the scientific-technical ideal of control and management continues to win the battle against the other pole of the cultural dialectic—that is, the ideal of personal freedom—can be attributed to the fact that the former ideal makes use of the *objective* cultural powers, which manifest themselves in new scientific and technical possibilities such as system theory, computer science, computer technical applications, and genetic manipulation techniques. Moreover, economic powers strengthen that process. However much the criticism of that process is increasing, a far-reaching cultural revolution is actually almost impossible. The reason for this is to be found in the existence of economic powers which have unbounded ambitions and in the life of the masses of consumers who continue to support the present primary direction of our culture, because they believe in and hope for still more material benefits from science and technical applications.

The threat of the current dialectic

It is necessary to emphasize that in this historical process the cultural dynamic is more and more assuming a malicious, life-threatening character. Modern technology and the use of its possibilities are reaching unheard of heights and are taking on a despotic character. By means of scientific-technical control and management of the entire world, not only are human beings' freedoms being curtailed, but raw materials across the whole earth are being exhausted, Nature is being rapidly destroyed, and the environment is being irreversibly polluted.

Recently a lot of attention has been given to the issue of climate change. An unbridled scientific-technical dynamic is making havoc of natural, ecological, energy, and social borders, whereby the outbreak of wars between nations is a real possibility (Van der Wal, et al., 2006, 223).

In developing countries, because of the influence of globalizing technical and economic developments, strong feelings of political powerlessness are growing, exacerbated by continuing economic malaise and stagnation. This is experienced as a direct humiliation. In other words, the scientific-technical culture of the West puts other cultures under pressure, via globalization. The dialectic manifests itself as a conflict between cultures, peoples, and nations. Through it, cultural tensions can discharge explosively and violent political conflicts can erupt.

The new dimension of the present-day configuration of the cultural dialectic has two components. Until now, resistance remained—as we have seen—limited to *subjective* resistance. Because people didn't have objective, cultural power at their disposal, that resistance could not be realized in terms of a change of the "technological culture," at most only adaptation was possible. The first thing which is genuinely new is that the resistance to the foreign "technological culture"—coming from Islam—is now a reality. It nestles in Western culture, and, at the same time—and that is the second thing which is new—makes use use is made of *objective* cultural power. Terrorism is a true threat (Gray, 2007).

A Western philosopher such as Waskow—a revolutionary utopian—could call us to topple our technical culture by violence in the 1960's (Waskow, 1968). But he could reach no more than a verbal revolution. The present-day terrorists have considerable cultural power and extensive technical possibilities at their disposal and form a world-wide network by means of modern technology—think of the Internet—against which they are protesting, ironically enough. The destruction of the Twin Towers in New York in 2001 makes it clear that they are capable of destroying one kind of technology by another kind. It is this development which is so disturbing.

How do Muslim ideologues react to the current cultural situation?

Criticism from Islamic intellectuals

Sayyid Qutb (Buruma, 2004, 36, 116 f., 124 f. 131), one of the most influential Islamic, Egyptian thinkers of the 20th century, defended pure Islamic community against a growing Americanism that is conceived as the empty, idolatrous materialism of the West. In the course of his life, he became more and more embittered toward the behavior of the West, and therefore, opposed every possible assimilation. Just as all dreams of total purity are, his ideal of spiritual community was a fantasy which contained the seeds of violence and destruction. He is the founder of the Islamist ideology which carried on a confrontation with the most important ideologies of the West. He responded to Western arrogance with Islamic intolerance (Huntington, 2001, 333). The radical purity of Islam and the destruction of the West are his aims. Thereby, Qutb has become a representative of a radical Islam which doesn't shrink back from violence in its resistance to the West. Instead, he even recommends it explicitly (Qutb, 1990). With him, the cultural dialectic becomes the fuming motor of destruction.

Thankfully, there are also Islamic reformists who seek a harmonious society, such as Mohammed Iqbal. This Pakistani thinker is not enamored of the Occident and expresses criticism of the West from his Muslim perspective. He focuses, in particular, on the visible, unbridled scientific-technical developments in the world, on the financial power of Western capitalism, on the economic exploitations inherent to capitalism, and on the secularism which accompanies it, or is even caused by it. He criticizes the influence of the West, because, in his view, humanity is being pulled away from Allah by the European Enlightenment. Now the West serves idols in His place. Iqbal therefore criticizes Western arrogance, imperialism, and public morality. But in doing so, he does not take a distance from science and technology (Buruma, 2004, 122, 152). On the contrary, he takes the Allah's Unity—so well-known to Muslims—as the basis for his thinking about science and technology. According to Iqbal, that unity must be reflected in human society by harmony that is expressed as justice, equality, solidarity, and care for Nature and the environment. With this he shows his support—in accordance with the spirit of early Islam—for reforms of science and technology (Iqbal, 1971; see also Foltz, 2003). He would like to help reduce the extant cultural tensions.

Points of contact in Islam

The Pakistani Muslim and Nobel Prize winner for physics, Mohammed Abdus Salam, argues in the same spirit for an acceptance of technology. In a lecture entitled *Science and Technology in the Islamic world* (Salam, 1983), he says that Allah has given everything in heaven and earth to humanity in order to be made use of.

In Salam's view, the Muslim scientist must seek to gain insight into the world, and by doing so, into Allah's plan. Science must be an integral of the human community in order to further material welfare. That's why Salam aims at the universality of science and technology. In order to achieve success in this realm, humans must be grateful to Allah and, as a consequence, obey Allah's will.

Salam wants to return to the first age of Islam, the period in which the torch of scientific and technical development was passed from generation to generation in order to reappreciate the proper motives for the blossoming and development of science and technology. For Salam, Islam is thus *necessary* in order to have the proper motivation and ethics for science and technology. By expressing himself in this way, this Muslim scholar has spoken about the relation between or interplay of religion and technology in a way which is new in the present-day Muslim world, and which in Western Enlightenment thinking occurs too little or even not at all.

Criticism of technology from Christian philosophy

We observe that Western culture is judged quite one-sidedly by reformist representatives of Islam. It is historically justifiable to maintain that the Enlightenment has Christian roots. But this cultural movement has pulled itself more and more away from Christianity and now opposes Christianity repeatedly. That's why it is incorrect for Islam not to make much of a distinction between the influence of Christianity and that of the Enlightenment, as if both necessarily would lead to the same ethics for technology (Buruma, 2004). This, while it is orthodox Christianity, makes a trenchant critique of the Enlightenment dialectic, as I have shown.

In the course of the 20th century, both ideals of the Enlightenment—the ideal of personal freedom and the scientific-technical ideal of control and management—have landed us in crisis situations which have had disastrous effects on global culture. This internal Western cultural struggle between freedom and scientific control is increasing. And now a radical and violent Islam is becoming stronger and stronger in its opposition to the West. In other words, Western culture is being more and more internally undermined and externally threatened.

No one less than Habermas (2005)—originally a thoroughly Enlightenment philosopher—has recently shown that a "wrecked Enlightenment" needs religion. Huntington says in his book about the clash of civilizations that the clashes between Islamic and Western culture are above all a result of the weakening of Christianity as the central component of Western culture (Huntington, 2001, 335). The question immediately arises; can a culture, with the loss of its religious roots, survive (Hittinger, 1995)? A renewal of Western culture would mean that people return to the religious source of Christianity and that Christianity understand its cultural calling and follow it.

A renewed Christianity would, on the basis of a powerful faith conviction, call for a modern transformation of Western culture. This call would be responded to from many sources. Already we hear voices calling for change. I think of the work of the Roman Catholic theologian Küng, who writes of coming to a world ethos in science and politics, "a global ethics" (Küng, 1997). As well, world-wide church organizations have already produced reports with criticism of the development of Western culture (Opschoor, 2007).

There is something very valuable in these appeals. Yet, in my opinion, in this analysis our social and cultural problems are attributed too much to dysfunctional, exploitive, *economic* relationships. There is not enough insight into the fact that the ideals of the Enlightenment, which concentrate on freedom, science, and technology, form the basis of the great cultural problems and tensions of our day. The Enlightenment ideals are, in fact, contradictory to each other. How can this contradiction be resolved? Instead of the idea of an autonomous freedom, free from all norms, we must discover and promote a freedom which corresponds to values such as order, discipline, authority, respect, trust, mutual assistance, and human solidarity. That is, a freedom which is intimately bound up with responsibility.

A renewal of the motives for science and technology is also needed. The quest to exercise dominating power should be replaced by an attitude which seeks to *serve* in the perspective of *world-wide justice*. The values and norms for technology must no longer be borrowed from the technical worldview, which finally leads nowhere. This insight is needed precisely because technology is foundational for so many cultural activities.

It is clear that it is now quite automatic to seek technological solutions to the knotty problems which technology itself has caused, which in turn, predictably, become new problems and threats. That's why another vision of technology is needed in order to diminish and even find solutions to these kind of problems. The erratic, high flight of technology needs a firm, transcendent anchoring. How can this be achieved?

Acknowledging God as the Origin of creation and humanity as the responsible image of God that has a divine calling to disclose and unfold reality as God's creation, makes the purpose of science and technology subordinate to the divine purpose of the history of the Kingdom of God (Sweeringen, 2007, 271 f.). Instead of the primacy of the technical worldview leading us forward, as the Enlightenment posited, the disclosure and unfolding of creation as God's Garden that is destined to become His glorious Garden city, should become primary and beckon us onward (Schuurman, 2005). In short, with respect to the transformation of "technological culture" which is so longed for, Christianity opposes the "religion of the material" just as much as reformist Islam.

It is evident that, thankfully, there are also many outside of Christianity and Islam who realize that fundamental changes in Western culture are needed. They are needed not only because of the inward threats coming from within Western culture itself but also due to the threats which come from the outside, such as radical Islam. Thereby extra support can be expected for a necessary paradigm shift within "technological culture" coming from the ethics of a reformist Islam (Hassan, 2007, Soroush, 2007), in its concern to care for Nature and the environment, and for social justice—in spite of the remaining big differences with Christianity.

Kuhn's paradigm theory

In order to have a better idea of what such a paradigm shift might mean, I would like to use Kuhn's paradigm theory regarding the development of science. Kuhn has made clear that the rise of scientific theories can be explained in sociological, psychological, economic, and even religious terms. By doing so, not only is the continual growth of scientific knowledge explained, but the unexpected "leaps" within science are clarified. The continual development of science demonstrates stability and agreement between scientists. In the case there is crisis in a field of scientific development, this leads to adopting a new framework—the paradigm—within which that scientific field can develop further. Only when the new paradigm emerges and triumphs can the rupture around the crisis be healed and a new period for that scientific field proceed. This means that, in the light of such a paradigm shift, the absolute truth claims of science are significantly relativized (Kuhn, 1962).

Kuhn teaches us that, in the midst of crises of scientific theory forming, suddenly big, fundamental questions are asked. The old scientific faith in certain theories shakes on its foundations. What were once commonly accepted ideas are no longer so. The once close-knit community of scientists in a particular field begins to dissolve into factions. Unanimity crumbles away. The "tacit knowledge" of the like-minded begins to waver. In short, the old paradigm begins to be seen as outmoded. A new theory emerges and begins to hold sway (Koningsveld, 2006, 110 f.).

Could Kuhn's vision of the shift of paradigms in science be able to function as a model for a necessary change in our paradigm for culture? An analogous model teaches us something useful but has its limitations. For example, science is merely a branch or part of culture. Culture contains so much more than science. But precisely because our culture is more and more seen as "technological culture," or "scientific-technical culture," we are stimulated to nevertheless be inspired by Kuhn.

The transformation of "technological culture"

Isn't something like this—a relativizing of the existing cultural paradigm and its transformation—possible in our conception of the current cultural developments? Within the dominating cultural paradigm of the West, we are faced with many problems. At the moment we're trying to solve these problems with the same means and methods which have actually caused these problems in the first place. Solutions are evidently part of the problems of our culture, particularly when support is sought from the economy and from politics. Slowly we are starting to realize that this is no longer possible. Is there a chance that we, in the midst of this crisis, might be able to find a way forward towards a new phase of culture wherein the problems of "technological culture" can truly be addressed and begun to be solved?

A cultural revolution or cultural about-face, analogous to a scientific revolution, will be necessarily accompanied by tense discussions, which finally must lead us back to assess what people believe and see as being true. Here the role of religion comes in view. Different critiques of culture and technology are being expressed from the standpoint of religion or religions, as we see coming from Christianity and a reforming Islam. The challenge is to see if another cultural paradigm can be formulated. This could limit or even begin to solve the existing problems and threats. That is not easy, for representatives of the old cultural model do not give up easily. They hold on to it, sometimes with the persistence of pit-bull terriers. These opposing forces have an economic, political, and cultural character. But at the same time, as the current developments continue to increase, we see more and more the weaknesses of the old paradigm. Isn't it true that we see an increase of world-threatening events as a result of the current scientific, technical, economic way of thinking?

The conflict between industrial and biological agriculture

I believe that there are possibilities for change. One concrete, strikingly relevant example of cultural change, both in the West and in the Muslim world, is the fight which organic agriculture—with or without success, and with or without sufficient arguments—is carrying on against industrial agriculture (Foltz, 2003, 3 f.; Petruccioli, 2003, 499 f.; Schuurman, 2005, 49 f.). Industrial agriculture is the cause of more and more problems. From the side of industrial agriculture a growing number of opponents of organic farming are making their voices heard, while the supporters of organic agriculture are becoming more vocal as well.

As we see the vague contours of a new paradigm begin to take shape, more successes are being achieved. Conversely, the defenders of industrial agriculture are calling for an ecological approach to agriculture. Both developments show that the existing problems are being taken seriously, and that people are looking for new, sustainable paths to take (Simons, 2007, 240 f., 340 f.).

A cultural about-face

A comparable about-face ought to take place in the whole of "technological culture." The political and economic worlds are opening their eyes to cultural alternatives, sustainable development, and socially responsible enterprises. Our social-economic climate is becoming more favorable to drastic changes. For example, we see how recent reports coming from the business world and aimed at the political arena, are calling for more attention to be given to the environment and climate change (Willems, et al., 2007). As well, a recent United Nations report about climate, coming from a world-wide scientific cooperative project of some 2,500 scientists, labeled human beings, and their technical applications, economies, and consumption as the most important cause of the huge emissions of greenhouse gases, with all the risks that entails (Report of the IPCC, UN Climate Committee, 2007).

As attention is being given to climate change, rising ocean levels, the shift of climate zones, the disturbance of ecological systems, the loss of biodiversity, and new tropical sicknesses, an appeal is being made for a change in our cultural ethos. Bill Clinton and Al Gore have also been calling for changes in these areas. We must not underestimate the influence, in the course of the years, of the Greenpeace organization. More and more people are opening their eyes and discovering a new cultural paradigm. Increasingly, people are seeing that our modern society, with its patterns of producing, controlling, and consuming, is inherently, intentionally, non-sustainable (Van der Wal, et al., 2006, 8 f.).

The existing, old patterns of culture are being undermined by all of this. Political parties in The Netherlands are looking seriously at the idea of sustainability. At the same time,the dominating cultural elites are eliciting more and more doubts about their own character as an establishment. Political forces may eventually work in a positive way to change the cultural disposition of many.

When, in addition, the consumer also begins to get insight into what sustainability concretely means and that the quality of our life is to be benefited by new measures, the conditions are favorable for real cultural change. I do believe that a very necessary cultural about-face is about to happen. And this will hopefully include more attention given to world-wide justice, over against the injustice of the current globalizing developments.

It is thus very important that our post-industrial Western culture attempt to address the threats of industrial culture. It will be a necessary learning process. I think that the increased interest in religion has everything to do with this scenario. Fundamental questions are being brought forward by religious spokesmen after a long period of being disregarded. What is the essence, the purpose and meaning of human life, of culture, of technology, and of the economy? Starting with these fundamental questions—stemming from the religious roots of cultures—lines are being drawn to touch all the different sectors of life which shape culture.

Analogously to Kuhn's terminology, we can speak here of a "gestalt switch," a "turnabout," or "revolution." This means a sudden "leap." Correctly so, for it is "time to turn." We can conclude that, in terms of our Western cultural history, the cultural experiment of the Enlightenment has turned out to be a large-scale failure—however much we may give thanks for its benefits. A radical turnabout of our culture is called for. A new orientation, a new, meta-historical compass, is needed!

The content of a new cultural paradigm

What does the new cultural paradigm look like? What is its essence? It must be substantially different from the previous paradigm, while still absorbing and preserving that which is of value in the old paradigm.

In the old cultural paradigm, Nature is seen as lifeless and mechanical, and within that framework, exploited by means of boundless manipulation. Nature, humanity, the environment, plants, and animals were, in accordance with the technical paradigm, looked at from a technical perspective—the so-called "machine model."

Change is now the order of the day. Now the protection of life itself must become the all-determining objective in the formation of culture. Science, technology, and economy must not be allowed to destroy life in all its variety and richness, but rather they must find their reason for existence in serving life.

However much Christianity and reformist Islam differ from each other—and I must stress how impossible it is to finally combine them with each other—at the same time they have much in common. This enables them to recognize each other as partners, participating in the new cultural about-face (Rohrmoser, 2006). For both Christianity and Islam, the model of the Garden is appropriate to this end (Petruccioli, 2003, 499 f.; Schuurman, 2005, 37 f.). Together they can agree with the following statement, "We love the whole creation for the sake of the Creator" (Foltz, 2003, 29). The Christian and the Islamic religions can, each in their own way, contribute to a globalizing culture in which life is no longer threatened, but enriched, and in which more genuine justice is put into practice, so that tensions are reduced. In spite of all the major mutual differences, there will be more social cohesion and mutual peace.

If that is not successful—for example, because of a lack of faith vitality in Christianity and/or lack of support from reformist Islam—the struggle between the ambitions of the Enlightenment and radical Islam will intensify and Islamic violence increase. Then there will indeed be reason for pessimism with respect to the future (Bawer, 2006).

Summarizing conclusion

In industrial society, technical thinking is still dominant. Virtually everything is viewed in the light of the technical model or the machine model. In this model, life itself is missing as a fundamental, decisive factor. The power of technology is worked out in a despotic and destructive way on the basis of this model.

Granted, in the new cultural phase we cannot and will not do away with much of technology itself. But it is clear that technology must function as a servant to life and to society. Reality can no longer be regarded as merely an object of technical manipulation, but must be seen as a creation and a gift of God, which we ought to accept with love and gratitude. This demands from us awe for the Owner, receptivity, modesty, humility, a sense of wonder, reverence, and prudence.

This new approach to technology will be characterized not by destructiveness as a result of seeking domination, but by seeking to carefully disclose and unfold reality, however much they may differ. Christianity and reformist Islam cherish the perspective of the living Garden city, shown us in Genesis 2. In this perspective there is appropriate care for the environment and Nature. The preservation and well-being of life on earth become thus more important than material prosperity.

This new framework for culture, by which life and love become the grounding categories, also seeks to strengthen justice as a mandate, and seeks to honor the supra-subjective, normative boundaries in creation that are established by God. Hereby it becomes possible to reduce cultural tensions and begin to realize a balanced, sustainable, peaceful, and richly varied development. An attitude of moderation will also help to reduce cultural tensions and threats, within both Western culture, and also in relation to Islamic culture. Reformist Islam, when we see its ethical priorities, must be able to be won for such a cultural about-face.

One big consequence is that to the degree radical, violent Muslims refuse to go along with this scenario, they must be isolated politically, and prevented from obtaining the cultural powers of science, technology, and economy, including financial funds, subsidies, and weapons.

I believe that, in this concrete way of following this new paradigm, a more sustainable, just, healthy, and peaceful globalizing development is possible.

Literature

Al-Ansari, Abd Al-Hamid, 2007, *The Root of Terrorism is the Culture of Hate*, www.memri.org/bin/opener_latest,cgi?ID=SD162507.

Barbour, Ian, 1990, *Religion in an Age of Science*, San Francisco: Harper.

Bawer, Bruce, 2007, *While Europe Slept*, New York: Broadway Books.

Van Bommel, Abdulwahid, 2002, "Islamitisch wijsgerig denken," in *Cultuurfilosofie*, p. 295-340, edited by Edith Brugmans, Open Universiteit Nederland, Budel: Damon.

Buruma, Ian and Margalit, Avishai, 2004, *Occidentalism*, New York: Penguin.

Daiber, Hans, "Die Technik im Islam," in: Stöklein, Ansgar und Rassem, in Mohammed, *Technik und Religion*, Düsseldorf: VDI Verlag, p. 102-117.

Dooyeweerd, Herman, 1959, *Vernieuwing en bezinning-Om het reformatorisch Grondmotief*, Zutphen: J.B. van den Brink & Co.

Foltz, Richard C., Denny, Frederick M. and Baharuddin, Azizan (eds.), 2003, *Islam and Ecology—A Bestowed Trust*, Cambridge, Massachusetts: Harvard University Press.

Gray, John, 2007, *Black Mass: Apocalyptic Religion and the Death of Utopia*, New York: Farrar, Straus & Giroux.

Habermas, Jürgen, 2005, *Zwischen Naturalismus und Religion*, Frankfurt am Main: Suhrkamp Verlag.

Al-Hassan, Ahmad Y.,2001, "Factors Behind the Decline of Islamic Science after the Sixteenth Century," Epilogue to *Science and Technology in Islam*, Part II, UNESCO.

Hittinger, Russel,1995, "Christopher Dawson's Insights: Can a Culture Survive the Loss of Religious Roots?" in *Christianity and Western Civilization*. Ft. Collins, CO: Ignatius Press.

Hoodbhoy, Pervez, 2007, "Science and the Islamic World—The Quest for Rapprochement," p. 49-55, in *Physics Today*, August.

Huntington, Samuel, 1996, *The Clash of Civilizations and the Remaking of World Order*, New York: Simon & Schuster.

Iqbal, Mohammad, 1971, *The Reconstruction of Religious Thought in Islam*, Lahore: Shaikl Muhammad Ashraf.

Koningsveld, Herman, 2006, *Het verschijnsel wetenschap*, Boom: Amsterdam.

Küng, Hans, *Weltethos für Weltpolitik und Weltwirtschaft*, Wissenschaftliche buchgesellschaft Darmstadt, München: Piper Verlag GmbH.

Kuhn, Thomas S., *The Structure of Scientific Revolutions*, Chicago: University of Chicago Press.

Newman, Jay, 1997, *Religion and Technology—A Study in the Philosophy of Culture*, London: Praeger.

Noble, David F., 1997, *The Religion of Technology—The Divinity of Man and the Spirit of Invention*, New York: Alfred A. Knopf.

Opschoor, Hans, J.B., 2007, "Wealth of Nations or a 'Common Future': Religion-based Responses to Unsustainability and Globalisation," p. 247-281, in: Klein Goldewijk, Berma, ed., *Religion: International Relations and Development Cooperation*, Wageningen: Academic Publishers.

Petruccioli, Attilio, 2003, "Nature in Islamic Urbanism: The Garden in Practice and in Metaphor," in Richard C. Foltz, Frederick M. Denny and Azizan Baharuddin (eds.), *Islam and Ecology: A Bestowed Trust*, p. 499-511 Cambridge, Massachusetts: Harvard University Press.

Qutb, Sayyid, 1990, *Milestones*, Indianapolis: American Trust.

Rohrmoser, Günter, 2006, *Islam—die unverstandene Herausforderung—Kurz kommentar*. Bietigheim: Gesellschaft für Kulturwissenschaft.

Sadri, Mahmoud and Sadri, Ahmad (eds.), 2000, *Reason, Freedom & Democracy in Islam—Essential Writings of 'Abdolkarim Soroush*, Oxford University Press.

Salam, Mohammed Abdus, 1983, *Science and Technology in the Islamic World*, Keynote Address delivered at the Science and Technology Conference, Islamabad.

Schuurman, Egbert, 1973, 1980 (2nd ed.), "De spanning tussen technocratie en revolutie, " 1973, in: *Techniek: middel of Moloch*, Kampen: Kok.

Schuurman, Egbert, 2003, (Vriend, John, translator) *Faith and Hope in Technology*, Toronto: Clements Publishing.

Schuurman, Egbert, 2005, *The Technological World Picture and an Ethics of Responsibility*, Sioux Center: Dordt College Press.

Simons, Petrus, 2007, *Tilling the Good Earth—The Impact of Technicism and Economism on Agriculture*, Potchefstroom, South Africa: North-West University Press.

Soroush, Abdolkarim, 2000, *Reason, Freedom, and Democracy in Islam*,

Oxford University Press.

Soroush, Abdolkarim, 2004, *Ethics and Ethical Critiques*, www.drsoroush.com/en/category/interviews/.

Soroush, Abdolkarim, 2007, *Dialogue of Cultures instead of Dialogue of Civilizations*, www.drsoroush.com/en/.

Stöcklein, Ansgar and Rassem, Mohammed (publishers), 1999, *Technik und religion*, Düsseldorf:VDI Verlag.

Swearengen, Jack Clayton, 2007, *Beyond Paradise—Technology and the Kingdom of God*, Oregon: Wipf & Stock Publishers.

Van der Wal, Koo and Goudzwaard, Bob, (eds.), 2006, *Van grenzen weten— Aanzetten tot een nieuw denken over duurzaamheid*, Budel: Uitgeverij Damon.

Willems, Rein, et al., December, 2006. *Pleidooi voor een kabinet met een mondiale visie op natuur- en klimaatbehoud*, Open Brief aan de leiders van de politieke partijen in de Tweede Kamer der Staten Generaal, Den Haag.

Waskow, Arthur, I., 1968, *Creating the Future in the Present*, in: Futurist, Vol. 2, nr. 4, WWR (Wetenschappelijke Raad voor het Regeringsbeleid), 2006, *Dynamiek in islamitisch activisme—Aanknopingspunten voor democratisering en mensenrechten*, Amsterdam University Press.

Zayd, N. Abu, 2006, *Reformation of Islamic Thought: A Critical Historical Analysis*, WRR-verkenningen, nr. 10.

CHAPTER 6

THE TRANSFORMATION OF MATERIALISTIC CULTURE: TECHNOLOGY[1]

Introduction

In our time among Christians, there is a lot of talk about the relationship between Christian faith and science, and about the relationship between that faith and economic development. Much less attention is given to the significance of Christian faith for technology, but it is precisely in our own time that reflection about faith and technology is so needed.

Our age is one experiencing many crises—financial, economic, ecological, the crises around energy and raw materials, all deriving from a fundamentally moral crisis. The remarkable fact is that people who analyze these crises often give much attention—and rightly so—to the negative results of (neo-liberal) economic trends, but scarcely give any attention to the influence of technology on economic reality and the significance of technology for diagnosing problems and seeking solutions to them. It seems that a kind of selective blindness is continuing in this area.

1. Technology as a cultural power

Everyone in society has to do with the reigning cultural powers such as the economy, national and local organizations, science, and technology. These powers provide the dynamic within culture. The basis of these powers, in my opinion, is technology; for without technology the other powers couldn't function.

What is technology?

Technology is fundamental for human culture and therefore indispensable. When we speak about human history, we often indicate the periods or ages with the aid of the level of technology: the stone age, the

---

1 Expanded special lecture for the student organization, Civitas Studiosorum Reformatorum of Delft University of Technology, 2013

bronze age, the atomic age, and the computer age.

After the industrial revolution, and especially since the appearance of the computer, we have entered into a completely new phase of history. Until this time, you could speak of particular technology, such as building houses, making copper kettles or wooden shoes, and building dikes. These old forms of craft technology remained pretty much the same, and were passed on from one generation to another. These forms of technology were just part of society in general, for they were not seen as existing on a higher level than other cultural phenomena. The old technologies were intertwined with what Nature provided—think of the windmill— even when they were deployed to combat threats coming from Nature, such as floods.

Since the appearance of the industrial revolution, but especially after the Second World War, technology has changed drastically. Science now has a much greater impact. The basic characteristics of science, such as rationality and universality, have become characteristics of technology too.

Modern forms of technology have left their distinct mark on society and culture. Interpretation is called for at this juncture. Modern technology is often seen as merely a neutral means of reaching certain goals. However, modern technology is much more than a means or a set of tools. You can't say "I'm making use of a computer or a car just the way I want to." For immediately we realize that, in order to drive a car a whole technically structured environment is needed, and in order to use a computer, you also need a technical network.

The French philosopher Jacques Ellul (1980) has written convincingly about the modern *phenomenon of Technology*, which he spells with a capital T. "Technology," with a capital, refers, for Ellul, to a totality of technical applications that are mutually connected to a whole, a system, or a network. This is something different than anything which has historically belonged to technology up till now. Earlier forms of technology were discrete phenomena, separate from each other. They were also culturally a more or less peripheral phenomena. Now, in our time, everything is connected to each other and these mutually connected technical applications work interactively. This modern Technology (as an interconnected system) lies at the very center of our culture. The interconnected form of modern technology leaves its mark on everything. In the philosophy of culture, the terms "Technological Culture" and the "Technological Age" (with capitals) are used, and correctly so. Later in this chapter, we'll see how this system of technology is connected to economic and organizing systems, and that science is the logical and functional connecting factor between the systems which enables them to cohere.

Autonomous Technology?

Ellul adds something else here. He says that Technology (as an inter-connected system) is autonomous. This means that, while people make a contribution to this system, they are actually subject to it. The Techno-logical-system process is, according to Ellul, wholly determined and not subject to human power. Ellul interprets this as something negative. He does this sometimes so forcefully that he can label modern Technology the greatest evil of our society.

In my opinion, Ellul's slant here is misleading and even dangerous. For if our Technology systems are really completely determined or ful-ly autonomous, then they become an unstoppable force in themselves, usurping all of life. As a result, you often hear people say, "You can't stop technology."

Here I believe that criticism is called for. Technology (with a capital T) remains the product of human effort, and human beings are respon-sible for their actions. If Technology was ethically or morally neutral, then human responsibility would be limited to the aims involved. Some-times engineers try to avoid responsibility for their products by saying: "engineers like us can't lie, that's what politicians and economists do!" (Dessauer, 1956). This giving-the-blame-to-others is very common. It was done, for example, by engineers who worked in developing nuclear energy. They claim, *they* were not responsible for the results, but the *pol-iticians and the economists* were!

The reason that many interpret Technology systems as "autono-mous" is that they are becoming more and more complex, and are being extended continuously, at an increasing tempo. We see an acceleration of expanding innovations. Many people are contributing to this state of affairs. They have no overview of the whole picture any more, and they can't keep up intellectually or ethically with the accelerating pace. The system of Technology is increasingly experienced as a maze in which humanity is trapped.

I would say that the technical development around us certainly gives the impression of being autonomous. However, people are still capable of intervening in the way systems are functioning. Sometimes, instead of accelerating, you must slow down the possible development at hand, so that justice is done to the human dimension of things. This means that human beings remain responsible for their actions.

It is true; responsibility is sometimes more difficult to pinpoint or profile than in the classical technologies of craftsmen. In the case of Technology (as vast interlocking systems), there is certainly a common responsibility here, not just an individual one. Common responsibility is increasing in its significance and is often more difficult to define. As well, due to the presence of different ideological convictions, there is often a lack of experienced, common ethics. Yet, however great the practical problems may be, we must continue to insist that every technology, and even the huge, interlocking systems of Technology of our day, is the product of human ingenuity, design, and industry, and that human beings therefore bear a final and clear responsibility—individually and jointly—for them.

Is Technology ethically neutral?

Many are of the opinion that technology and Technology systems are ethically neutral. The idea here is that technology can be *used* for good or bad purposes, but technology itself is neither good nor bad. This view is often held tacitly. But technology is never neutral. Think, for example, of the impact of technologies on Nature, in the case of the supplies of raw materials. And it is clear how the position of humanity has been radically changed by technology.

In the old, classical technical applications, technical applications can be seen as extensions of our own hands, our own five senses, and other parts of our bodies. In modern technology it's quite different; human beings are made subordinate to it. As well, ethical questions are posed regarding whether something is legitimate or not, and under what conditions. And if economic development has a detrimental effect on the culture or people, then it is legitimate to ask if applications of technology have enabled this kind of harmful economic development. Greed is often connected to economic motives, but Technology systems, as the basic structures of economic development, can stimulate greed in all kinds of subtle ways.

Technical applications are always a bearer of culture or bound to a particular culture. Just as culture is *value-laden*, so too is technology. Technology cannot be separated from worldviews, by which human beings understand themselves, have a conception of the world and of Nature around them, not to mention of God and the spiritual world.

Technology is not ethically or ideologically neutral because technology always entails or implies certain concrete applications, with certain contours. And in modern Technology systems there are factors and powers present which proceed in a definite direction, and *only* in that direction. Through the influence of the Enlightenment, technology has developed one-sidedly, in ethical terms. There is a correspondence or a deep relationship between certain human tendencies or motives and the development of technology. That means that it is not accidental that modern technology has developed in Western culture. Apparently Western human beings have held to certain values and norms which have motivated them and have satisfied deep needs. But which values are involved here?

How technology is viewed

In my doctoral dissertation, *Techniek en Toekomst* (1972) (English translation: *Technology and the Future* (2009)), I discussed in detail and criticized two dominating views, the views of the transcendentalists and the positivists. The first stream of thought represents the pessimists—hostile to technology—and the second the optimists—believers in the benefits of technology.

You could say that until 1980 these two streams of thought were the most significant ones reflecting philosophically about technology. Philosophers of technology speak about an *"empirical turn"* at this point in time, a turning to the practical side. The philosophy of technology became more descriptive, instead of expressing a judgment or view about technical developments. The philosophers of the empirical turn, in their ethical approach to special problems, concentrate on the practical aspects of technology, the so-called "cases," which they derive from the totality of technical developments.

In fact, the philosophers of technology have been seduced here into concentrating on trivia. An analysis of a particular case of a technological application has only limited value. It teaches us, for example, what went well or not well in a particular situation. These philosophers do not give much attention to the structural and "normative" development of technology as a whole. This is my objection to this approach. It seems clear to me that there is a coherence in the development of technology. In other words, an integral view of the phenomenon of technology is missing in the theories of the "empirical turn." And an integral view is precisely what is needed to embark on a viable, responsible path toward a deeper understanding of technology in the future.

More recently we notice technology philosophers speaking of going *"beyond the empirical turn."* The responsibility of the engineer and of society as a whole, including in the realm of politics, is happily once again a matter of discussion.

The dominance of technology in modern culture

It is quite common to see political and social responsibility as only being connected with developments in the economy. There are legitimate reasons for this. It certainly is striking that the power of money appears to dominate the economy and to disrupt it in the process. "The love of money is a root of all kinds of evil" (1 Timothy 6:10), the Bible says.

However, without doubt, there are more values in human existence than those connected to our attitude to "money." Here we can learn from Ellul. He says that the central place which Marx gave to the money economy in his social criticism no longer is sufficient to adequately analyze our own age, in the face of the phenomenon of the modern Technological system.

Technology—according to Ellul (1979)—and not the economy, is the definitive factor or motor in the development of our Western culture. I would say; both of them, *in a process of continual interaction*, are determining our culture. I do concur with Ellul that many people, surprisingly, have no awareness of the deep influence of technology.

Ellul says by way of explanation that it is not capital which stimulates technology, but technology which dominates capital. That is, modern technology, made possible by science, precedes and shapes capital and the economy. The focus of this is to be found, in our day, in computers; our wide-ranging, interconnected, information and communication systems. And this process is strengthened by the ever increasing possibilities of computer techniques. These technical applications must be seen in the totality of their effects and results. Applying them has far-reaching consequences for the economy and for the environment, and is one of the chief causes of the crises of our times. It is precisely when we are gripped by this power—controlling and managing power (technically) or the power of money (economically)—that we become blind to the negative effects or damaging results of the phenomenon Technology. In a certain sense, we don't have adequate antennas to recognize how *treacherous* Technology can be. We think profits can be made through technical applications, so the latter are good things. However, we are more often than not the victims of an illusion.

Technology dominates the economy

In general we can say that Technology forms the basis for economic development. Without Technology as a foundation, our modern economy is unthinkable. Technology and the economy, together, have produced much that is positive. But if we only see the positive side, we don't see the reverse side of the situation.

The first example. Maarten Schinkel (in the Dutch daily NRC of Jan. 6, 2012) points in a column to a blind spot in the analysis of many cultural crises. Until now, he writes, no one has signaled the generally disastrous influence of technology in their critical views of culture.[2] Schinkel shows clearly that through the influence of technology "the ecology of the currency market"—as he calls it—has been fundamentally altered. By means of involved, complex technology, we no longer have insight into what is actually happening. Could it be that the banking crisis of 2008, involving the disastrous sub-prime loans, demonstrates in fact the "betrayal" of (unbridled) technology? Bank products which

---

2   Schinkel uses the term technology, which, as I often emphasize, is actually the *science* of technical applications, as so many do. But I acquiesce to current usage, and use the term in both senses, depending on the context.

had been constructed by computers—the so-called derivatives—were no longer transparent. Even the then president of the Dutch National Bank, Wellink—and many beside him—had to admit that he didn't know exactly what they had sold with these bank products. Technology had masked the actual value of such products. People regarded them as being valuable, but eventually it became evident that they had no value whatever. You could say that by means of technology, money was manufactured out of thin air—an impossibility.

In the course of Schinkel's piece it becomes clear how great the threat of technology can become to the value of money, and therefore to the economy as a whole. He suggests that internationally connected, world-wide computer systems ought to be set up in order to be able to understand what is happening in the world-wide financial markets.

However, the question immediately arises: Is he suggesting to solve the problems of technology by another set of technological solutions? That is comparable to wanting to "cast the devil out by the devil." For if national or European computer systems are not able totally opaque, why would international systems do better? The example of Schinkel's column shows us that we need to unmask the "blessing" of technology as such. The hidden, malevolent ideology of technology must be brought to light.

Don't misunderstand me here. I'm not saying that computer systems must be thrown out of the banking world. What is needed, at least, is that the developers of such systems must be ready to indicate what criteria they are using when they are constructing these systems. They must make clear what they are doing and what they are responsible for.

The second example. In searching for a solution to the recent crisis in Europe, the decoupling of certain European countries from the European Union was categorically rejected. Different economically "sick" countries, such as Greece, cannot exchange the euro for their original currency. The economies of Europe are so tightly bound to each other that the "sick" members cannot be isolated from the rest. It is evident that this mutual dependency has been created by the technologies involved. And no one is able to really penetrate to the bottom of it or get an overview of it.

The third example is possibly even more convincing. Martin Ford (2010), in his important book about automation, robotics, and the economy of the future, has illumined the problem of unemployment in this time of mounting crises. He wants to describe these crises not in financial terms—as others generally do. It is very clear, he says, that the financial problems of the American banks were the initial causes of the crises, but that we have not noticed sufficiently that an exploding increase of technology is the hidden factor here. This is evident when we look at the growing unemployment in the world as a whole. It is also clear that this unemployment is structural in character, caused by accelerating technology such as information systems, nanotechnology, 3-D printers, robotics, intelligent computer "brains," self-driving cars, etc.

In spite of the crises, and in the face of growing unemployment, the automatic production of goods and services, via technology, is continuing to increase. This is happening—in contrast to what is commonly maintained—in both developed and underdeveloped countries. Financial solutions for the crises alone will not be able to reduce this unemployment. We must give much more attention to the growing and omnipresent influence of technology if we genuinely want to prevent permanently dysfunctional societies.[3]

Blind to the dangers

As we have seen, modern technology causes great changes in society, as it forges an increasing alliance between many technical applications in the areas of information and communication, genetic engineering, nanotechnology, neurotechnology, and other areas. These new technologies, connected to the economy, are the most significant determinant of social change in our time. It is this social role of technology which presents us with the greatest challenge. If technology is so significant for society, how can it be developed in such a way that it can truly contribute to solving the great social problems of the 21st century?

In the past I personally have opposed the idea that the economy is the only really significant motor of social developments. The difference between the beginning of the 20th century and the beginning of the 21st century is huge, and this has become clearly evident with the advent

---

3 These facts ought to be very important for political policy. Through the policy of singling out the "top sectors" for attention, the Dutch government has been trying to escape from the current crisis by means of technology and economic measures. Politicians want to promote economic growth as the way to combat the financial problems.

But it could very well be that even when this is successful, unemployment will continue to increase, perhaps less so for technicians and those with higher educations, but especially for those who are school dropouts or those with a low-level education, it will be difficult to find work, due to the new technical developments, robotics, and self-driving cars.

Think of people working in clothing factories, taxi-drivers, and cashiers at checkout counters. But even those working in white-collar jobs, such as telemarketers, insurance agents, and tax advisers, will find themselves threatened with unemployment. In the past, new technology created more jobs than it eliminated. Now that perspective will be considerably more difficult to achieve, because current technologies are signaling a wave of renewal, not just for one or a few sectors of the economy, but for almost the entire economy, and that at a super-fast tempo, whereby it is difficult to develop new jobs quickly, or to have people be retrained for the new situation. The working middle class is diminishing fast. And we cannot see any plans of the government for making this a top priority for the future. For more information about this area, see Martin Ford, http://econfuture. wordpress.com/2010/04/06/did-advancing-technology-contribute-to-the-financial-crisis?

of the new technologies. These technical changes have, together with ideological, social, economic, and political changes, deeply colored the 20th century. I find, however, it more and more surprising that the often disrupting influence of technology is not being signaled, even in views which are critical of our culture—including those expressed by Christians. People have apparently gotten used to seeing a failing economy as the cause of all problems. People have developed a blind spot and therefore do not see how technology plays a crucial role in what is actually taking place.

The assumption that technology is autonomous or ethically neutral helps confirm people in not wanting to look critically at it. Furthermore, the technological developments are increasing the tempo of our life so fast that we are less able to be empathetically sensitive in perceiving the moral problems of our age. How can we be healed of this moral blindness (Bauman, 2013), and how can we encourage true social renewal?

We must learn to see that the positive aspects of technology are being overshadowed by the negative ones. Who would ever deny that technology has made great contributions to material prosperity, education, housing, and health? But it is also clear that the negative influence of technology is very evident in the crises we are facing. This is visible in the areas of the environment, climate change, the growing scarcity of raw materials, providing a sufficient food supply for people, and also the dangers of nuclear energy. The impossibility of getting an overview of the complex developments of technology makes it even more difficult.

Further, we may not forget the problems of health care and the loss of social cohesion and social trust, because individualism and the loss of social values and norms go together with the rise of modern technology.

The sociologist/philosopher Ulrich Beck (1986) says that the chief characteristic of the (post)industrial society is that, with advancing technology, the risk of catastrophic dangers has greatly increased. Furthermore, the attempt to take "precautionary measures ahead of time" and to give attention to "sustainability" has not led to sufficient changes (Beck, 2007).

A basic cause of this is that people, in spite of the negative developments, continue to think positively, in a naïve way, about technology. People are held captive by the myth of technological progress, by which all criticism of technology has become a taboo.

When criticism *is* expressed, people almost automatically assume that the critic is rejecting technology as a whole. This is something I have personally experienced many times, even while I was doing my best to emphasize that I see technology as gift from God, even a God-given mandate in being involved with creation.

Technological disasters a recurring theme

Terribly destructive, even apocalyptic disasters have taken place in our time, underscoring the *presumptuousness* of modern technology. Think of the nuclear disasters of Chernobyl and Fukushima, the oil spill disaster in the Gulf of Mexico or, years before, the disaster at the chemical plant in Bhopal, India. These horrific, large-scale events, causing unimaginable suffering, are illustrations of what is potentially taking place in the whole world, under the leadership of Western culture. They belong, intrinsically, to our current culture. In modern technology there is a hidden, destructive potential of unprecedented dimensions. Yet who is signaling this?

We are no longer in control. We have become so obsessed with the new technological possibilities, promising more material prosperity, and giving us new forms of amusement, that we have become anaesthetized to the dangers involved. We are perhaps still masters and control the realm of small technological developments, but in terms of the large-scale technologies, they *holds us hostage*!

It is striking that brilliant thinkers, such as Einstein and Heisenberg, have expressed their deep doubts about the direction of modern scientific-technical developments. Einstein feared a "degeneration of technical society." Heisenberg compared technical culture with a "super-tanker," which in the long run would threaten to go out of control. Many of us recognize that feeling, but don't know what to do with it. What does the future hold?

What is happening? The world in which we live, especially since the rise of the computer and the connection of the computer to many other technical applications, has become a thoroughly *technical* world. Modern culture has been shaped through and through by technology. Technology has become the dominant force in world history. Without technology, we would not have a global economy. The totality of reality has been taken over by technology, and all human goals have become technological goals.

Since the time of the Enlightenment, the reigning opinion has been that technical progress will naturally go together with material prosperity. Technology is, on the one hand, the product of human beings, but the reverse is also becoming true; we human beings are increasingly adapting ourselves to the omnipresent technological world all around us. The result is that we don't really see what is happening and what the enormous dangers are. The norm for technical development is becoming technology itself. The idea is not that is possible, should be put into practice.

According to the German philosopher Ursula Meyer (2006), human beings are becoming "heimatlos" (homeless); we are losing the sense of having a meaningful purpose in life, and we experience morality as a maze, or a whirlpool. And this is even more serious because it happens without our being conscious of it. Technology is penetrating everyone's thinking and way of life without our awareness.

Let me give an example. Technical thinking fosters certain behavior patterns that have the effect of reducing the totality of life. The technical thought pattern which is built into computers influences the uncritical user. He or she will mentally follow, more and more, the same pattern, over and over. When we human beings are functioning day and night by interacting almost exclusively with technology, the result is that our minds and energy are becoming possessed, as it were, by technology.

Our age is the age of "social media," and many people have "friends" via these media, but the friendships are often mainly technical in character. How many friends on Facebook are ready and willing to show real friendship in case of real need?[4] Looked at in this way, the "social media" often have, ironically, great asocial effects.

It is clear that technology is now not simply a means to reach certain goals, but is rather an impersonal, cultural power with far-reaching effects that extends its tentacles into all we do and are. We cannot maintain that technology automatically produces prosperity, freedom, and happiness. There are positive effects, that is certain; but the negative effects are too often underestimated and too often underreported. The danger of losing oneself in the technical world and the threat to our human future is real.

How then can we avoid the latter scenario? How can the dangers, threats, chaos, and potential disasters be prevented?

Culture paradigm

When we look at the history of technology, we do well to look at the motives, values, and norms of its development. Nevertheless, I prefer an approach which more directly interprets the unity and coherence of technical development in relation to science and the economy.

I've tried to discover the dominant culture paradigm which has risen in Western culture, and which, in my opinion, ought to be replaced by another paradigm. A cultural paradigm shift—an ethical framework within which culture develops—is called for. This is something I have been emphasizing in many lectures over the past years.

There are all kinds of developments going on in our time—think for example of striving for "sustainability," applying the principle of "taking precautionary measures ahead of time," and developing a "green economy"—which can be seen as positive. However, what is missing is an all-inclusive approach. Such a paradigm approach provides insight into motives, values, and norms of the past, and also gives attention to nec-

---

4   Concerning this problem, see Sherry Turkle's book, *Alone Together* (2010).

essary fundamental values and norms for the future. Hereby we enter into the territory of ethics. Via the approach of seeing what a cultural paradigm shift might mean, we achieve a deeper and clearer insight into what we need to have a truly hopeful future.

A technical culture model

Since the beginning of our modern age, with as "father" the philosopher René Descartes, Western thinking has become more and more a *technical* way of thinking, seeking control and management of reality. Descartes makes use of a technical rationality in order to solve old and new problems of humanity and culture, and in order to increase material prosperity. Descartes says that the laws of mechanics are the same as the laws of Nature. He sees Nature as a totality made up of automatons. In other words, Nature is subsumed under the category of mechanisms. Hereby the natural sciences in particular are used as instruments, with the claim that everything can be made subservient to human beings. Hereby we see the mechanizing of the (Western) worldview (in the terminology of Dijksterhuis (1996)) making a revolutionary entrance in history. "Nature is a machine, as simple to understand as clocks and automatons, when we investigate it closely enough," Descartes says.

The consequence of this way of thinking is the idea that Nature can be mathematically calculated and managed. Because for Descartes humanity is the "maître et possesseur de la Nature" (the Lord and possessor of Nature), human beings are seen as being able to control and manage Nature and make it subservient to them.

Thus, following this Cartesian style of thinking, everything in Western culture is looked at through the spectacles of what is technically possible—on the basis of mathematics and the natural sciences. Since the Enlightenment, *the cultural model* has become a *technical* model. New technical applications are accepted as interpretation models of reality. For example, technical (computer) models dominate the economy and—as we saw earlier—the financial world.

We also see technical thinking and this technical interpretation of reality clearly present in the concepts of what it means to be human. An example is the currently popular and much-discussed study of the neurologist Dick Swaab (2010) of the University of Amsterdam: "*We are Our Brain*." The brains of human beings—as determining everything else we do—are interpreted according to the example of modern information and communication technology. In such a conceptualization it becomes clear how abstract the technical model is, and that applying it to everything does no justice to the coherence and fullness of reality. This model reduces reality, and herein lies the source of many problems.

In line with this kind of thinking is also the suggestion of the futurologist Ray Kurzwell (2013) to "download" to a computer the complete knowledge of the working of the human brain, whereby, according to Kurzwell, eternal life can come within reach. For the machine—the computer—will function, after this download of the operation of the human brain, in essentially the same way as our brains. This is an enormous overestimation of technical possibilities, to put it mildly. This comparison of a "dead"—inorganic—thing, which the computer really is, with *eternal* life is an unbelievable mistake, and testifies to a completely false view of technology.

This kind of technical thinking plays a great role, as well, in the currently popular discussions about human beings and hyper-modern technical applications, such as computers, robots, nanotechnology, etc. This discussion, led by "transhumanists," I won't discuss here. But I want to make this remark: Technical thinking about the connection between human beings and the most modern technical applications, which sometimes goes as far as to claim that in the future we won't be able to see the difference between human beings and machines, has, as its consequence, that human beings are no longer seen as being *responsible* beings.

To summarize, the technical paradigm has become and is still predominant in giving shape, in Western culture, to understanding, controlling, and renewing reality. This reality is understood technically from the word go, and we are seen as being capable to further shape this technical reality by means of our latest scientific-technical discoveries and inventions, and naturally—as is thought—improve it in this way.

Reductionisms in the technical culture model

In this development and in the concepts which lie behind it, we have to do with a thoroughly technically interpreted and controllable, material reality. In this *material* reality, humanity is Lord and Master, and our ethics is a *technical* ethics, with effectiveness and efficiency as the chief norms. This vision of humanity is the result of a materialistic self-delusion, a kind of spiritual eclipse. According to this view, we derive the values and norms of our actions as human beings from technical possibilities above all. Reflecting on this situation, the sociologist/philosopher Bauman speaks in his latest book (2013) of moral blindness present here. It might be better to say that, in this conception, ethics and morality undergo a fatal, reductionist, narrowing of vision.

Recently this narrowing of vision has become more than obvious in a book about the possibilities of nanotechnology. K. Eric Drexler—who himself was one of the fathers of nanotechnology—has written a book with the title *Radical Abundance: How a Revolution in Nanotechnology Will Change Civilization* (2013). He sees the future as a kind of technical paradise. He is totally unconcerned about the influence of reductionism in his concept of reality, and ignores the realities of evil and war.

Looking at things from a Christian, biblical point of view, we see here in the developments I've sketched cases of enormous reductionisms of reality. In the technical worldview, there is no divine, transcendent reality. Immanent reality is reduced to the world of a technical worldview. The latter contains a flattened, material reality in which all divine secrets have been removed. Nature has thus become a technical reality which we can improve.

Marx already said that Nature in itself is meaningless and purposeless, and that *we* give meaning and purpose to it by means of technical applications. Thus a godless world is reduced to a material reality, which is fully technical and controllable. In this process, humanity itself arrogates to itself the position of Lord and Master, choosing its own technical will and power as ethical guides.

Remarkably enough, this humanity and its culture are more and more the victims of what we have brought into being. Life in all its variety is threatened.

That a purely technical approach to reality—stimulated by economic profit motives—can be catastrophic, has become evident from the results of the greatest oil spill catastrophe in history in the Gulf of Mexico. In the borehole, in what was called the Macondo field, of the production platform Deepwater Horizon, an explosion took place, whereby the technical installation caught fire and sunk. Beyond the loss of human life, the life of the waters of the Gulf was severely threatened, and in many cases destroyed, in the course of three months. An area the size of The Netherlands was completely polluted by oil. With every new storm, oil was released into the environment from the ocean bed. Long-lasting pollution was the inevitable result. The fisheries and the tourist industry of the area were forced to shut down indefinitely. In order to repair the damage, 50 billion dollars were needed, and there was and is still great uncertainty about the future of the area. Since this catastrophe, many genetically malformed shrimp and crabs have been found in the ocean there. In short, the dominance of the technical culture model is *deadly*![5]

II. The transformation of the technical culture model

The biblical culture model: the Garden as desired cultural paradigm

In the old culture paradigm, Nature is seen as lifeless, as "dead"—inorganic—and within that framework Nature is exploited by unbounded manipulation. Therefore, while up till now Nature, the environment, plants, animals, and humanity itself was looked at from a technical per-

---

5   See the Dutch daily NRC, Feb. 25, 2013, p. 24,25

spective—the so-called "machine model"—through the technical paradigm, *now* is the time to protect life in cultural formation, and make this a perspective which governs all our thinking and activity. Science, technical applications, and the economy will not be allowed to destroy life in all its variety and richness of forms and expressions, but rather will be deployed in service to life. Operating from this perspective, technology and the economy will be more able to fulfill their goals.

A Christian view of reality—proceeding from faith in God as Creator and Sustainer of this world—is the acknowledgment that we live in a *created* reality. Everything in the universe is from, through, and unto God. Human beings are God's image, and love to God and our neighbor are the central guiding principles of everything for which we bear responsibility. The future is God's future in His Kingdom. Because of the Fall into sin, we live in a mangled, lacerated reality. There is coming a moment in time in which all of this will be done away with, when God Himself, through Christ, will establish His Kingdom. That is a glorious Kingdom, filled with divine glory and infinitely wide and deep, in which the citizens of that Kingdom will live in a never-ending state of harmony and gratitude.

If we think about this in a deep way, it's actually amazing that Christians do so little with the implications of this Christian vision. Perhaps we are still more influenced by the ideas of the Enlightenment than we know. A new "Enlightenment" by God's Word is therefore needed!

This Christian, biblical view of history implies that we adopt a completely different culture model than the technical model. The technical model is abstract and is a model characterized by "death." No wonder that following this model on a grand scale has led to reductionisms of reality, immense problems, and fatal threats. There is much too little attention given to life as it really is and to the fullness, the coherence, and the concreteness of reality.

When, instead, attention is given to a culture model in which life is in view, then technology is given its legitimate place and catastrophes caused by technology can be reduced. I believe that the old biblical culture model of the "Garden in development" fulfills the requirements I've enumerated. History is then seen proceeding from paradise as an unfolding Garden (Genesis 2) to the "Garden city of the future" (Revelation 21). Through the Fall into sin, this original perspective was enormously disrupted. "Thorns and thistles" and death characterize the path of human history. But this vision remains God's plan for history, giving us a tremendous, beckoning perspective. Through the crucifixion and resurrection of Christ, there is once again a hopeful perspective of the coming Kingdom of God. In Christ, all of our culture will appear as a Garden City beyond our immediate horizon at the end of history. Acknowledging this purpose and meaning to history—human thinking, instead of being domination thinking, becomes thankfulness thinking, and therefore seeks to be active in service. This gives a genuinely new direction to technology. The cosmological, anthropological, and ethical reductionisms and thus deficiencies in the current materialistic culture are uncovered and potentially overcome in this approach.

In the development of the Garden, life keeps its central place in spite of the unruliness and stubborn limitations which reality retains in our situation. Technology—old and new—is called to serve the life of humanity, animals, and plants. In the image of the Garden, life is not threatened, but enriched, and more real justice can be achieved so that tensions are lessened and social cohesion, wholeness, and peace is promoted.

The transformation of materialistic culture

Technology in service to life, ought to be the motto of engineers, politicians, and economists, all of whom have constantly and intensively worked with technology in our age. As a result, the continuous call for technical innovations, in connection with the growing economic problems, ought to be evaluated by the question, "what are these innovations actually for?" Innovations for the sake of innovations simply means a confirmation of the technical model of cultural development.

I see this happening in the case of nanotechnology. Take the book, mentioned earlier, by Drexler (2013) and the book of Diamendis (2012). Promoting nanotechnology as a given is to reverse the proper priorities. Nanotechnology is certainly promising, but what should it be used for? Evaluating innovations—and so also the possibilities of nanotechnology—by the culture model of the Garden poses the central question as to how any particular technology is to serve life.

Why are good reforms so difficult? First, because of the result of deeply ingrained historical patterns, western thinking in our day is above all technical thinking. Against this background, a complex of cultural powers of science, technology, the economy, and organization has been built up in time. Through the dominating influence of an instrumental science on the other cultural sectors, this complex demonstrates an abstract scientific character. It lacks the concreteness of real life, and has no awareness of our full and coherent reality as fundamental and definitive facts. Working from the basic perspective of a dominant materialism—which in the course of time has become a fundamental, religious attitude for many—attention is paid almost exclusively to the strength—purely in terms of material benefits—of that complex. And, indeed, that complex has brought much material prosperity and has developed—partly under the influence of economic concentrations of power—into a world-wide system.

But the reverse side of this—its weak side—has become more and more clear with the increase of crises and threats which typify our age. If we seek to solve the problems we're facing simply by means of the existing technical complex I've described, then we arrive at a point where not much more than about 10% of this complex is really sustainable in the long term.

If, on the other hand, we seek to mobilize this complex to be a means of service of everything that lives, then the abstract complex of cultural powers must not be seen as a world which exists in itself. No, the cultural powers must be brought into relationship with the fullness of created reality in its deep inner cohesion. In that fullness of reality, we then are able to see the blessing of material prosperity, but also the destructive influence of it on human beings and the environment. An *instrumental rationality* must be transformed into a *serving rationality*. In technology, preserving and caring must certainly receive just as much attention as constructing and building. In the economy, it must not be a matter of merely raising productivity, with the norms of effectiveness and efficiency, but instead be geared to seeking a healthy giving of care to our human world of living and working.

From a renewed biblical culture model we gain, over against the technical culture model, a long-range perspective. Reality as it is can no longer be regarded as an object of technical manipulation, but is seen as a given—as creation and gift of God—to be received with gratitude and love. We are called to have awe for the Owner of this world, along with receptivity, modesty, humility, a sense of wonder, respect, and prudence. Not lording over, but having respect for living reality in all its many colors and variety, and knowing that we are called to show love toward the world-wide community of humanity, give us another way of looking at technical development. Not destroying by controlling, but opening up and unfolding reality must be the aim of technology.

Such a Christian view of the unfolding of the creation gives us the perspective of the living Garden city. The threats and the problems we face can be addressed within visible boundaries. We will be more able to reduce cultural tensions and threats in Western culture itself but also in relation to non-western cultures. A more sustainable globalizing development will be possible, and world-wide problems and threats can be pushed back. Certainly, a technical paradise will not be the result if we proceed in this way. But genuine progress is possible, with God's help.

Following this Garden model does not mean that we can do away with existing technology. All the possibilities, technical discoveries, and innovations are built into God's creation, even now when the creation is suffering from severe disruptions due to the Fall. That's why we must value the calls to sustainability, which we hear from all sides at the moment. This is so even when non-Christians are the authors of these calls; they are sometimes the first to emphasize the importance of sustainability.

Here we glimpse something of God's common grace, even in the midst of the wreckage caused by the technical culture model. God in His care and love for His creation prevents the ship of humanity from going too far in a potentially disastrous direction. And further, "sustainability" should not be isolated as a concept, but needs a continual extension and deepening of meaning, provided by a renewed cultural model.

On the basis of human creativity and our multidimensional responsibility, engineers are called to devote themselves to serving, protecting, enriching, and strengthening all living things. Technological applications, developed from the model of the Garden, ought to correspond to the Great Commandment of love for God, but that also implies appreciation of cultural values, social values—such as solidarity, the common interest, justice, valuing people's innate worth, care, and charitable works—as well as ecological values, political values, and economic values. Thus we see how more than material values are to be acknowledged. That is a demand of justice itself. This kind of a cultural renewal contains within itself the desire and hope of beginning to solve the crises of our time.

Technology must therefore do justice to everything which has value, as the philosopher Brey (2008), from the University of Twente in The Netherlands, has beautifully stated. This starts with the small things in life. We can think of concentric circles which gradually increase in size and reach. In our day, the whole world has become the last, big circle. Thereby it is appropriate, for example, to give attention to the climate crisis and the crisis of raw materials, which threatens all forms of life world-wide.

Criticism of technology strengthens criticism of neo-liberal economics

In the beginning of this chapter I said that much of current culture criticism devotes itself exclusively to criticism of our economic order. I share that criticism. However, it must be supplemented with criticism of the technical world. The views of well-known critics of neo-liberal economics, such as Goudzwaard, Graafland, and Jongeneel, would be stronger, more challenging, and also more successful if they would, in their presentations, take account of the fact that the powers which are active in technology need to be dethroned just as much as the powers of money in our modern economies. The "dead" technical computer models of the economy need to be integrated with the biblical Garden-model, so that the economy is enabled to serve everything which lives, and at least everything which is vulnerable. Justice and mercy ought to be the leading motives in caring for Nature and the environment and in seeking the growth of social and human capital. Attention must be given to a healthy and health-giving care of our world—as Goudzwaard (2009, and in other publications) supports—that is, of our entire human world of living and working. In my view, giving another direction to technology is necessarily a correlate to the "comprehensive approach" as favored by Goudzwaard.

The image of the Garden-in-development is clearly connected to the original meaning of "economos." Giving care, cherishing, guarding, protecting, and preserving go hand in hand with cultivating the ground, harvesting, and producing. What we need to see, instead of the current increasing scale and tempo of production, is this new paradigm of the Garden put into practice; leading us to strive for a scale and a tempo which favor a humanity in partnership with the rest of God's creation. In this image of the Garden, the limits of Nature's ability to bear burdens are to be respected. The idea of making beneficial use of the fruits of the Garden implies the direction we need to go to achieve a more sustainable development of culture.

In a recent study of the Rathenau Institution[6] an appeal is made, correctly, to support a fully organic economy. A next step ought to be to define "sustainability" better. Sustainability involves not just the guarantee that the needs of future generations be met, but also involves the need to protect and guard the realm of plants and animals. This demands wisdom and prudent stewardship. Then it will be possible, for example, to prevent the production of bio-fuels from replacing the raising of crops for food. True sustainability represents the cycle of the rhythm of life and the rhythm of the organic. This does not exclude culture, but includes it fully. The relationship between technology and Nature ought not to be one of exploitation through the domination of technology, but rather Nature and technology ought to harmoniously go together in a cooperative fashion.

In this way there will be more attention given to justice, over the continuing injustices of the current globalizing developments. Sustainability is possible, operating from the metaphor of the Garden, that is, when technology, together with the economy, no longer results in manipulation, exploitation, and pollution, but, as the economist Herman Daly of the World Bank has expressed it, when they help maintain and increase the world's capacity for fruitfulness and production, and make sure that these bounties and products are available to all people, now and in the future.

Responsible cultural development means to live from the "interest of the capital" which God has given us, and refuses to allow that capital to be damaged or consumed. This is, in fact, the central notion involved in seeing humanity as a caretaker and steward. We can begin by concurring

---

6   Rathenau Institution, *Naar de kern van de bio-economie*, Den Haag, 2011.

with the present, urgent demand for sustainability and by supporting the thought that the economy must "serve the people, the planet, and profit," in that order. Further, the principle of "cradle to cradle" stewardship is necessary for industry so that a throw-away culture is exchanged for one which recycles raw materials. These examples, already partly being put into practice, demonstrate that we are, thankfully, already seeing a certain shift in the direction of this new, needed paradigm, away from the old and still dominant one.

Politics and technology

This approach to renewal and change is called for in the area of politics, as well. I could say a lot about this, but I would like to restrict myself to looking at Dutch national indicators; for example concerning the Dutch Gross National Product (GNP) and the reports of institutions, such as the Social Economic Board (SER), which pre-eminently belong to the old culture of the technical applications paradigm. They are dominated by technical thinking. Such evaluations and reports needn't be directly abolished; but it would be advisable if they could be supplemented by new indicators and institutions focusing on other issues.

Next to the SER, it would be very beneficial to have a *Social Council*, in which employers and employees, environmental groups, and development and safety organizations would participate on the basis of common responsibility and common interest. Next to the GNP as an indicator of our current technical, economic development, a more realistic "well-being indicator" is definitely called for, as an alternative. By means of such a new indicator, the cultural, social and human impact of policies, as well as the technical, ecological, and economic aspects of development, could be assessed and interpreted. But because such an indicator is difficult to measure and assess quantitatively, we could begin by establishing a national sustainability index, next to the GNP.

But there is more to be done in the area of politics. It is a common misunderstanding to think that everything in our society is driven by economic development. For behind and underneath every economic or financial activity in our day, we see technology as the ground structure. Changing the economy or financial policy can be considerably resisted by technological systems which stubbornly remain the same. The current economic developments are one-sidedly oriented to a neo-liberal, *laissez faire* approach. It ought to be different.

Political action is needed in order to develop a new, appropriate, fitting technological basic structure which is dedicated to serving the life of citizens and their social relationships. As well, our political struggle in connection with the financial crisis and the bank crisis can only really succeed when technology becomes more geared to human values. That's why transparent, publicly accessible computer programs should be demanded to be used in international and European financial transactions. This is needed to prevent uncontrolled concentrations of financial power, which always tend to make fraud and deception probable.

Suppose we want to make sure that motor vehicles don't go faster than 40 kilometers per hour in our villages and cities, to make things safer. But suppose four lane highways are allowed to be built, with the use of modern techniques, through these same villages and cities. The result will be that the speed limit is set much higher than 40 kilometers per hour, and the policy of achieving safe traffic will have been de facto abandoned.

Instead, in order to achieve such safe traffic, we have to adjust the size and shape of the roads involved, adding "speed bumps" if needed. Supporting, appropriate, fitting technical measures are taken, which make a safe traffic policy more effective.

Technology must demonstrably serve and not destroy or oppress.

Happily, we see some encouraging developments at the moment, including increasing attention given to social business enterprises, a call to reduce the exploitation of Nature and pollution, a movement encouraging ecological and organic agriculture, diminishing the bio-industry, growing intention to address the climate problems, and the development of green and sustainable energy. This indicates that a shift is visibly taking place in politics, as the old cultural model is making way for a new one.

The New Engineer

What is the significance is of the new cultural paradigm for the different technical studies at the university? As a veteran civil engineer I know of some beautiful examples of completed projects where technology has had a serving function.

One is the Oosterschelder dam, on the west coast of Zeeland, here in The Netherlands. This dam, by being able to be kept open to the ocean except during storms, prevents the body of water behind the dam from becoming a "dead sea." Much of the natural environment and the biodiversity of the Schelde river and estuary were able to be maintained and kept healthy through the use of this deliberately chosen technical construction method.

Another example is the ecological city neighborhood of Stockholm, Sweden. In this neighborhood, all cars are parked underground, and use is made, in part, of tunnels to leave the neighborhood. It is a beautiful neighborhood, which shows by example that a Garden city is already able to be achieved on a small scale.

And let's not forget the city of Delft, here in The Netherlands. When, 50 years ago, a massive railroad viaduct was erected, passing through the center of Delft, I agreed wholeheartedly with those who protested this ugly eyesore in this historic city and called for an underground tunnel to keep this beautiful city livable. At that time, this was impossible to achieve. Happily, today a tunnel is finally being constructed.

Of course it's much more expensive now than it would have been 50 years ago. But that is not really a problem. Operating from the still-dominant modern technical model, we often get projects done at too cheap a cost with damaging results. Later we get the real bill, when much more money is needed to compensate for a cheap initial solution, whereas spending more at the beginning would have actually been much more economical in the long run.

Another concrete example—I'm very surprised that many people protest vociferously against building windmills to produce electricity at sea, because, in their opinion, the horizon is ruined. That may be true to a degree. But why isn't there just as much protest against cars being parked everywhere in this country. They are totally spoiling all the once beautiful views in so many cities and towns, and no one is demanding that all the cars be parked in underground garages.

In fact, every engineer has the obligation to be resourceful and creative, within his own field, in order to help achieve a more livable world. We need more variety and less uniformity.

I can point again to the example used at the beginning of this chapter, concerning the big banks and their abuse of power. These banks, which are definitely needed for our society, ought to be not only reduced in size, but ought to be run by a smaller group of managers.

Mega-projects, in general, do not conform to the criterion of enhancing responsible human cultural life. The ideas of the philosopher Schumacher, himself a Christian, argued for small scale projects in the energy crisis of the 1970's and are still highly relevant. Books of his such as *Small is Beautiful* (1973) and *A Guide to the Perplexed* (1977) are sure to inspire every new generation of engineers, in a contemporary way, to follow with conviction and joy the biblical culture model I've sketched. Here we find a hopeful perspective for our created, though now disfigured, reality!

Literature

Bauman, Zygmut, (2013). *Moral Blindness: The Loss of Sensitivity in :iquid Modernity*. Cambridge: Polity Press.
Beck, Ulrich, (1986), *Risikogesellschaft*, Frankfurt: Suhrkamp Verlag.
Beck, Ulrich, (2007), *Weltrisikogesellschaft: Auf der Suche nach der verlorenen Sicherheit*. Frankfurt: Suhrkamp Verlag.
Brey, Philip, A.E., (2008), *Techniek en alles wat van waarde is*, University of Twente.

Dessauer, Friedrich, (1956), *Streit um die Technik*, Frankfurt: Knecht.

Diamendis, Peter H. and Kotler, Steven, (2012), Abundance: The future is better than you think, New York: Free Press.

Drexler, K. Eric, (2013), *Radical Abundance: How a Revolution in Nanotechnology Will Change Civilization*, New York: Public Affairs.

Dijksterhuis, Eduard J., (orig. ed. 1950), *De mechanisering van het wereldbeeld*, Amsterdam: Meulenhof.

Ellul, Jacques, (1980), *The Technological System*, New York: Continuum Press.

Ford, Martin, (2010), *The Lights in the Tunnel: Automation, Accelerating Technology, and the Economy of the Future*, U.S.A.: Acculant Publishing.

Goudzwaard, Bob, (2009), *Wegen van hoop in tijden van crisis*, Amsterdam: Buijten & Schipperheijn.

Kurzwell, Ray, (2012), *How to Create a Mind*, New York: Penguin Books.

Meijer, Ursula I., (2006), *Der Philosophische Blick auf die Technik*, Berlin: ein-FACH-verlag.

Schumacher, Ernst F., (1973), *Small Is Beautiful: A Study of Economics As If People*, London: Blond and Briggs; (1977), *A Guide to the Perplexed*, New York: Harper; (1980), *Good Work* (with Peter N. Gillingham), New York: Harper Collins.

Swaab, Dick F., (2013), *We zijn ons brein*, Amsterdam: Atlas Contact.

Turkle, Sherry, (2011), *Alone Together: Why We Expect More from Technology and Less from Each Other*, New York: Basic Books.

Recommended Literature

Barbour, Ian, (1990), *Religion in an Age of Science*, San Francisco: Harper.

Beck, Heinrich, (1979), *Kulturphilosophie der Technik*, Trier: H. Blumenberg.

Ellul, Jacques, (1965), *The Technological Society*, New York: Knopf.

Ellul, Jacques, (1990), *The Technological Bluff*, Geoffrey W. Bromiley, trans., Grand Rapids: Eerdmans.

Gray, John, (2007), *Black Mass: apocalyptic religion and the death of Utopia*, New York: Farrar, Straus, and Giroux.

Habermas, Jürgen, (2005), *Zwischen Naturalismus und Religion*, Frankfurt: Suhrkamp Verlag.

Hittinger, Russell, (1995), "Christopher Dawson's Insights: Can a Culture Survive the Loss of its Religious Roots?" in *Christianity and Western Civilization*, Ft. Collins, Colorado: Ignatius Press.

Massink, Henk, (2013), *Blijvend Thuis op Aarde*, Delft: Eburon.

Newman, Jay, (1997), *Religion and Technology: A Study in the Philosophy of Culture*, Westport, Conn.: Praeger.

Noble, David F., *The Religion of Technology: The Divinity of Man and the Spirit of Invention*, New York: Knopf.

Opschoor, Hans B., (2007), "Wealth of Nations or a 'common Future': Religion-based Responses to Unsustainability and Globalisation," in

Klein Goldewijk, B. ed., *Religion, International Relations and Development Cooperation*, Wageningen: Academic Publishers, p. 247-281.

Ratzinger, Joseph, (2007), *Values in a Time of Upheaval*, New York: Crossroads Publishing.

Schuurman, Derek C., (2013), *Shaping a Digital World: Faith, Culture, and Computer Technology*, Downers Grove, Illinois: InterVarsity Press.

Schuurman, Egbert, (2003), *Faith and Hope in Technology*, Toronto: Clements Publishing.

Schuurman, Egbert, (2002), *Bevrijding van het Technische Wereldbeeld: Uitdaging tot een andere ethiek*, Concluding Lecture, Delft University of Technology.

Schuurman, Egbert, (2007), *The technological World Picture and an Ethics of Responsibility*, Sioux Center, Iowa: Dordt College Press.

Simons, Petrus, (2007), *Tilling the Good Earth: The Impact of Technicism and Economism on Agriculture*, Potchefstroom: North West University Press.

Stöcklein, Ansgar und Rassem, Mohammed, eds., (1990), *Technik und Religion*, Düsseldorf: VDI Verlag.

Swearingen, Jack Clayton, (2007), *Beyond Paradise: Technology and the Kingdom of God*, Eugene, Oregon: Wipf and Stock.

Van der Wal, Koo en Goudzwaard, Bob, (2006), *Van grenzen weten: Aanzetten tot een nieuw denken over duurzaamheid*, Budel: Uitgeverij Damon.

Van der Walt, Bennie J., (2007), *Transforming Power: Challenging Contemporary Secular Society*, Potchefstroom: Institute for Contemporary Christianity in Africa.

Verkerk, Maarten, Hoogland, Jan, Van der Stoep, Jan, De Vries, Marc J., (2007), *Denken, ontwerpen, maken: Basisboek Techniekfilosofie*, Amsterdam/Meppel: Buijten & Schipperheijn en Uitgeverij Boom.

ABOUT THE AUTHOR

Egbert Schuurman (born in 1937 in Borger, The Netherlands) studied civil engineering at Delft University of Technology and philosophy at the Free University of Amsterdam. He received his doctorate in 1972 with a dissertation called *Techniek* and *Toekomst: Confrontatie met wijsgerige beschouwingen* (English translation: *Technology and the Future: a Philosophical Challenge* (2009)). From 1964 to 1966 he worked in the area of soil mechanics at Delft University of Technology, and in the area of cultural philosophy at the Free University of Amsterdam from 1966 to 1984. In 1972, he became special Professor of Christian Philosophy at Eindhoven University of Technology, and remained there till 2004. He had the same position from 1974 to 2004 at Delft University of Technology, and from 1984 to 2007 at Wageningen University. From 1983 to 2011, he was the chairman of the ChristenUnie (Christian Union) political party delegation in the First Chamber (Senate) of the Dutch Parliament. From 1983 to 1984, he participated in the international research team project concerning *Responsible Technology* in the U.S. From 1981 to 1983, he was a member of the so-called Broad DNA Committee, which, commissioned by the Dutch federal government, studied the social and ethical aspects of the use of human hereditary material. He was a an editor of two international magazines in the area of philosophy and technology. He was the member of the steering committee of the Dutch Royal Institute of Engineers which had the task of investigating *The Limits of Technical Applications*. Schuurman gave guest lectures on subjects related to the philosophy of technology in Canada, the United States, England, Korea, Japan, South Africa, France, and Brazil. In 1994 he was awarded the *Templeton Award* for teaching about *religion and science*, granted to him in Berkeley, California, at the Center for Theology and Natural Sciences. In May 2002, he gave a concluding lecture at Delft University of Technology with the title: *Liberated from a Technical Worldview: a challenge to frame new kind of ethics*. In 2003, he was named an Officer in the Order of Orange and Nassau. In September 2007, he gave a concluding lecture at Wageningen University with the title: *The challenge of the Islamic critique of technology*.

Next to his doctoral dissertation, *Techniek en Toekomst*, translated into English (*Technology and the Future*, in its 2nd printing) and into Chinese, he has published, among other works, the following: *Techniek: middel of Moloch?* ("Technical applications: means to an end or Moloch?"), T*ussen technische overmacht en menselijke onmacht: Verantwoordelijkheid in een technische maatschappij* ("Caught between technical power and human powerlessness: responsibility in a technical society"), *Christenen in Babel* ("Christians in Babel"), *Het technische Paradijs* ("The technical paradise"), *Filosofie van de Technische Wetenschappen* ("The philosophy of technical sciences"), *Geloven in Wetenschap en Techniek—Hoop voor de Toekomst?* (Faith and Hope in Technology, Toronto: Clements Publishing). The following books are being prepared for publication: *Reflection on Technology and on the Technological Culture* and a Spanish translation of *Faith and Hope in Technology*.

www.ingramcontent.com/pod-product-compliance
Lightning Source LLC
LaVergne TN
LVHW041255080426
835510LV00009B/743